£1.99
£4/43

Jeremy Clarkson's Hot 100

Virgin

First published in Great Britain in 1997 by
Virgin Books
an imprint of Virgin Publishing Ltd
332 Ladbroke Grove
London
W10 5AH

A catalogue record for this book is available from the British Library

ISBN 1 852277 300

Designed by Neal Townsend for J.M.P Ltd.

Printed in the UK by
Butler & Tanner Ltd.

contents

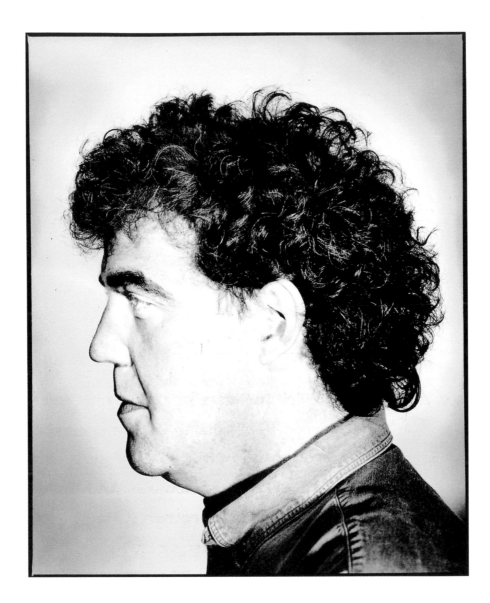

For Francie

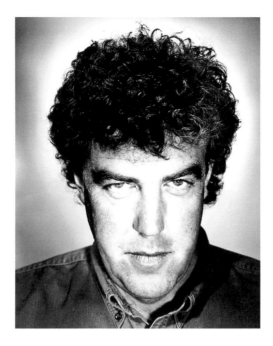

introduction

A car, as we all know, is merely a device for transporting some people and their luggage from point to point. Which means of course that my favourite car should be the Toyota Corolla which does the job asked of it better than anything else ever made. It never breaks down. The servicing is inexpensive. And the dealers always remember your children's birthdays. Aaaaah. Or rather, aaaargh.

You see, a car is rather more than the sum of its parts. To argue that a car is simply a means of conveyance is like arguing that Blenheim Palace is simply a house. That's why, in this book, you can search until your ears fall off but you won't find the Toyota Corolla. It is dull. It is tedious. It is a lowest common denominator car, built without passion. It is a Barratt home on wheels.

But you will also not find the Porsche 911 or even the Ferrari 456, both of which are built with enough passion to get the censors reaching for their scissors. They're not here because I don't really like them. I do, however, really like the Peugeot 504 estate and the Ford Transit van so that's why they're here, nestling among the Ferraris and Maseratis. I don't expect everyone to agree with my choices but I do hope you enjoy the book. It took me ages to write it.

AC Cobra

Hmmm- comfy Cobra!

This is the British way. A man with mad hair and a brown store coat, working until the early hours in his shed at the bottom of the garden, invents a new space-plane that can get to Australia from Heathrow in 12 minutes. He takes it to a large aeroplane making company where a middle management berk in pleblon trousers says 'Get out of my office. You don't even play golf and therefore, you cannot possibly be taken seriously.'

Eventually, a foreign company hears of the invention, and invites the inventor over for a chat. Sadly he's not a businessman so he readily agrees to sign over all rights for £30. And another great British invention is wasted. So it goes with the AC Ace – an underdeveloped little car from the 1950s which was powered by a two-litre engine that could

Alfa Romeo Alfa Sud

Forty years ago, Ferrari was nothing much to write home about while Alfa Romeo was an entire city full of precious jewels and elegant cathedrals.

Alfa had won every single racing event that the world could throw at it, except, I think, the 4.40 at Lingfield. But during the 1950s, at one motor race, they even had time to pull their cars in on the penultimate lap so they could be polished. That way, they'd look smart when they crossed the finishing line. But then, as was fashionable at the time, the Government stepped in, took them over and everything went to hell in a handcart. Alfas couldn't win a race because by lap four, they'd usually rusted away.

However, there was still a spark that refused to go out. In the 1970s, the Government decided that the southern half of the country needed some employment opportunities and instructed Alfa to design a small, volume car which could be built there. The Alfasud (AlfaSouth, if you like) was born. And while it was held together with spit and Kleenex, there is no doubt that few urban runarounds had ever offered a driver such a sublime driving experience. You could turn a Sud into any corner at any speed, and it would get you round. Oh it would slither and slide and there'd be the usual bout of front wheel drive understeer, but one thing was always constant. The driver would be grinning from ear to ear.

It was a cool car, too. At the time, no student at Sandhurst or Cirencester would be seen in anything else. That said, it should have been, but for some extraordinary reason wasn't, a hatchback. And if it rained in the night, you'd come out in a morning to find everything except the plastic steering wheel had been converted into a fine red powder.

Alfa may have forgotten how to make cars last, but the Sud proved they could still make them fast. A word of warning though. When Alfa tried to develop the Sud by fitting fatter tyres and bigger 1.5-litre engines, it lost its raw appeal. And when the 1.7-litre Sprint coupe came along in 1980, the Sud theme was almost completely ruined.

It may have looked good but to drive, the Sprint was about as nasty as a Spacehopper.**jc**

---

trace its roots back to 1919.

It was updated in 1957 when AC slotted a Bristol engine under the bonnet. This had been – er – taken from BMW after the war and it wasn't bad. Top speed went up to 117 mph. However, at around this time a larger-than-life American called Carroll Shelby, who had raced at one time for Ferrari, arrived on the scene, saying he wanted to put a big Ford V8 under the bonnet. And they let him. And the Cobra was born. Thanks to its massively flared wheel arches, it looked sensational but the first example, using a 260 cubic inch engine, was basically a thoroughbred pulling a bath chair. The handling was dreadful. However, in 1965, the suspension and steering were updated so they could handle the same massive 427 cubic inch Ford motor that was fitted in the GT40. This Mark III Cobra

developed 390 bhp and could rip up the road thanks to a colossal 480 ft lbs of torque.

As a result, 0 to 60 could be dealt with in just 4.2 seconds and if you had strong hair, the top speed was 165 mph ... in 1965, for heaven's sake. Obviously, that kind of grunt is tame by today's standards but the Cobra will never be out of anyone's Hot 100 because it is still the horniest-looking, two-seater sports car ever to have graced the planet. From any angle, it looks utterly menacing but let me tell you this. It is at its most intimidating when you're behind the wheel trying to put all that power onto the road. This is a car that, like its name sake, can bite. For the unwary, a Cobra can spoil your whole day.In fact, the wheeled

weight lifter has spoiled the last 30 years for millions. When news leaked out that someone had taken a tuned version to 170 mph on the M1, politicians went into a bit of a dither and promptly introduced the 70 mph speed limit.

A great car then, but with a nasty after-taste.**jc**

Alfa Romeo GTV6

Sounds perfect (in third gear at 60mph, in a tunnel with all the windows open)

I owned one of these once and it was, without any doubt at all, the single most unreliable piece of junk since Marx invented Communism.

I'd climb in every morning, start up the engine and then have to tap each of the dials in turn to coax some life out of them. Often they'd report major malfunctions but usually they were lying. So I'd set off wondering, not what time I'd get to work, but where, on route, I'd break down.

Sometimes, the engine would just stop as though it had suffered a catastrophic brain haemorrhage. I'd be cruising along, listening to the stereo with the windows down and the sunroof open and then with no fuss at all, the entire car would die. One day the linkage which connected the gear lever to the rear-mounted gearbox detached itself and flopped down onto the prop shaft, jamming it solid. The rear wheels locked solid causing smoke, brimstone and a noise they heard three time zones away.

Then there were the endless occasions when I'd depress the clutch to change gear and the pedal would stay down in the footwell, skulking like a scalded cat. There was also a phase when all the air would suddenly decide to leave the confines of the tyres. No kidding. I'd park it in a locked garage and in the morning all four tyres would be flat. Weird. And a damn nuisance to boot.

I haven't finished yet. The only reason the sunroof didn't break is because it wasn't electrical. It did leak though. As did the windows which were electrical and which therefore jammed. As did the mirrors. And the stereo.

And then there was the design itself. Here was a car with a raising tailgate, just like any modern sports hatch, but the rear seats didn't fold down so, practically speaking, it was worse than useless. And then there was the driving position. I'd love to meet the dummy on which the seat/pedal relationship was modelled because he can't be human.

So what, then, possessed me to buy a car that did almost everything wrong? And why, for heaven's sake, is it here in my list of 100 favourite cars? Simple. No, it was not especially nice to drive. Yes, the gearbox was at the back for perfect

music to my ears

Engine	.........sohc, V6, 2492cc
BHP	154
Brakes	..........discs all round
Top speed	...................125mph
0–60mph	..............8.4 seconds
Clutch	..............Russell Grant
When?	1981–1986
Today's cost	.....£3500–£5000

balance and there was a complex De Dion rear suspension but the steering was heavier than a photocopier, meaning that twisty, fun roads were best avoided.

The reason why it's here is two fold. First, its 2.5-litre V6 engine was, and still is, the finest-sounding propulsion device ever to be shoe horned under a bonnet. And second, the Bertone design was, and still is, just stunning to behold. Like Claudia Schiffer naked on a Maldives beach at sunset, only better somehow.

On its day, this wonderful-looking and glorious-sounding car could make you feel like a million dollars ... I think. I can't be sure because in the year that I had mine, it never really had its day. But after a year, it had its chips. I sold it to a Lebanese shop keeper.**jc**

Alfa Romeo GTV

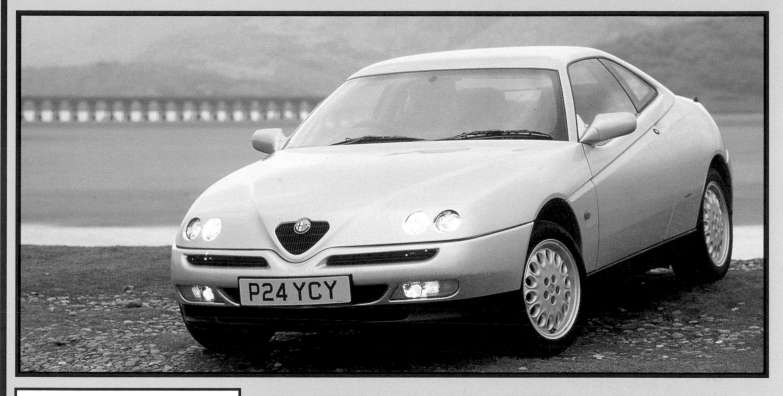

you're gorgeous

Engine	Straight 4, 1970cc.
BHP	150
Brakes	discs and ABS
Top speed	130 mph
0–60mph	8.4 seconds
Looks	Irresistible
Sounds	Pet Sounds
Today's cost	£23,500

So it's got a twin spark engine but come on; who cares? And the suspension doesn't really blow my frock up either. Yes, the steering is nice but I could list ten others cars, straight off the top of my head, where it's even better.

The interior is nothing much to write home about either though I quite like the way they've put the major dials in pods and I suppose things fall fairly readily to hand, but it's all let down by poor headroom and rather tacky plastic on the dashboard.

Dynamically then, it's a good car but when automotive historians are poking through the latter stages of the twentieth century, it won't even cause them to pause for breath. This is because it will have taken all their breath away, same as it does to me, every time I see one. Look through this book and tell me, hand on heart, that a more striking-looking car has ever been made.

Some, a very, very few, are its equal perhaps, but as a harmonious piece of design, it is a masterpiece. Look at the bold grille and the way those tiny, tiny headlamps bore out of the bonnet-like lasers. And the way its waist sweeps majestically upwards, giving the car a lean forward, purposeful stance.

Even the back works, which is something that can't be said about the GTV's convertible sister. This car is just one rung down from a Ferrari 355 and that means that when all is said and done, it's damn close to the top. **jc**

Looks lovely with Sean Connery as Bond in Goldfinger

Aston Martin DB5

See!

In its 85-year history, Aston Martin has built fewer cars than the city of Detroit churns out every three hours. And when it comes to making money, they'd have been better off standing outside a tube station with a wonky guitar and a can of Special Brew. Legend has it that there hasn't been a single year when Aston made a profit. Not once. Not ever.

The company has been kept alive by an endless parade of wealthy backers and by some outrageous good fortune.

The best bit of good luck came in 1965 when the Newport Pagnell factory got a call from Eon Productions, a little-known film company. They wanted to develop the James Bond character, and in his forthcoming film, Goldfinger, felt he should have a DB5.

At the time, Aston was in one of its bigger pickles. After 10 years of trying they'd won Le Mans and the World Sports Car Championship, but the assault had taken its toll. David Brown,

the owner at the time, needed a successful road car to pay off the racing debts. Then Bond took delivery of his DB5. As a result, it's one of the most instantly recognisable cars on the road, but rumour has it that we're not talking here about a particularly rewarding driver's machine.

Well, I drove one in 1996, and I thought it was a dream. Obviously, nostalgia plays a big part in the driving experience but I was impressed with its responsiveness – odd in a car that weighs about the same as an ice breaker.

It was fast too, although was pretty hard work. There was no air conditioning but you needed it because there was no power steering, either. And the brakes needed an almighty shove.

Make no mistake. This was not cutting edge technology, but the combination of Italian styling and British charm garnished with some Bondular glamour, puts the DB5 not only in the Hot 100 but very probably, in the Hot 10 too. **jc**

005

Engine	dohc, 3995cc
BHP	282
Brakes	discs all round
Top speed	141mph
0–60mph	8.1 seconds
MPG	15
Steering	hippo wrestling
How many?	886 (Volante 123)
When?	1963–1965
Today's cost	£35,000–£50,000

Aston Martin DB7

It'll get round most corners without crashing

When Ford bought Aston Martin in 1987, I really did think the world had ended. This was even more of a shock than the death of Keith Moon. Ford would not have a clue how to run a company like this. Ford would screw up. I was even making the Molotovs to attack Essex when the DB7 broke cover. And that, if anything, was an even bigger shock because it is – no arguments please – the best-looking car ever made.

It's funny but I've been buying albums and CDs for 25 years now and in all that time, I can only name one where I like, and I mean really like, every single track. It's Who's Next, in case you're interested.

Well the DB7 is the automotive equivalent of Pete Townshend's crowning achievement. Unlike any other car, there is no angle from which it is anything other than sensational. Even if you crawl underneath it and look up, your underpants will go all warm and gooey.

I'm sorry to bang on about this but some say the view from Kowloon across to Hong Kong island is the world's most spectacular sight but I'm not so sure. The DB7 is better. Frankly, if I had one, I'd put it in the sitting room and be happy to spend the rest of my life just gawping.

But this would be a shame because it's

actually quite good fun to drive. Powered by Jaguar's 3.2-litre straight-six engine, which for that little extra something, has a supercharger, it's plenty fast enough without being terrifying. And the same goes for its handling. We're not talking here about the rule books being rewritten because comfort was an important design parameter – but it'll get round most corners without crashing. Which is important.

There are a few problems. The seats are so big that rear visibility is impaired and life for bigger drivers is cramped. Plus, early models were screwed

together by people, who at the time, were obviously in a bad mood about something. However, if your DB7 breaks down, you should be glad because it means you have a chance to spend an hour or two ogling it while the RAC man comes.

Already, it's the best-selling Aston of all time and that's thanks in no small part to the world of football. In the past, those who have played with the round ball have had questionable taste but today's young lions are rather different. Ian Wright, David Platt and Ryan Giggs have all bought DB7s.**jc**

Aston Martin Vantage

When you climb into a Vantage, you'll find that the headrest has a squidgy cushion which nestles in the back of your neck. And as you pootle around, you'll find it a comforting sensation, like having a half-naked Fijian masseuse teasing away the stresses and strains of a working day. But it isn't there to relax you. It's there because this is the only car I know that can give an unwary driver whiplash, simply from the force of its acceleration.

Imagine sitting on a deck chair in your garden, half asleep on a lazy Sunday afternoon. Without warning a fully laden Boeing 747 doing 600 mph flies straight into the small of your back. Can you imagine that? Good, because now you know what it's like to dump the clutch in the most powerful road going Aston of them all.

The hand-built engine is not only colossal to behold but the power it delivers simply boggles the mind. 550 brake horsepower means its ten times more powerful than an original Golf GTi but that's only half the story. It develops

550 sq ft lbs of torque too which makes this twin supercharged V8 the most gruntworthy motor in current production. Of course, it needs to be because the Vantage will never win Slimming Magazine's Car of the Year award. It tips the scales at a huge two tons, making it even heavier than a night storage heater.

This has always been the Aston way. They need biblical power because no real thought is given to either weight saving or aerodynamics. The Vantage has the frontal area of a large house and inside, the appointments of an even larger palace. It really is a Rolls Royce with attitude.

The dash is fashioned from real wood and the leather really does look like it was once part of an animal. Too often these days, I'm presented with so-called leather seats that smell and look like they're nothing more than a petro chemical bi-product.

And in equal measure, I get to drive sports cars that have about as much to do with sporty motoring as cheese. The

Vantage, however, is different.

At first, it does feel like you're wrestling with a rocket-propelled chest of drawers. You dab the throttle and it lunges off in a relentless pursuit of the horizon, which is a bit un-nerving so you back off, and the nose dips like you've just crested the highest point of a roller coaster.

Now this is bad enough when you're going in a straight line, but things get a whole lot more scary when the first corner starts to fill your windscreen. What you have to do is stop, take a few deep breaths and remind yourself that no matter how imposing the car may be, you are the boss. It may be one of the most expensive, powerful and luxurious cars on the road but it's just scrap iron – and wood – without a driver.

So take it by the scruff of the neck – and it's like breaking in a wild mustang. The Vantage, please believe me, can be tamed but it only understands brutal treatment.

And then it becomes almost unbelievably good – like a super quick kart almost. There is understeer at the beginning of the corner but a little more power, or a little less, or a little more lock, or a little less, or almost anything really, will have the back coming round in a perfectly controlled bout of the most magnificent power oversteer.

I'm not an idiot. A Ferrari 550 does this better and so, for that matter, does a Nissan 200 but neither looks or feels quite so outrageous in the process. Getting a Vantage out of line is like skidding in Blenheim Palace. Only the Vantage is a little more uneconomical. On one serious bout of cross-country motoring with Elgar on the CD, I managed to get the fuel consumption down to 9 mpg, making it even thirstier than a 4.6-litre Range Rover in central London, on a Friday night. I love the way Vantages are made too. There's a man at the factory

who examines each body shell before it's painted and even the tinlest blemishes, completely invisible to the untrained eye, mean the whole damn panel has to be reshaped and recrafted. It takes three days to make a door and if that quality controller finds a hair's breadth mistake, the panel beater has to start from scratch.

And it's the same story with the engines which bear the maker's name in a plaque on one of the rocker covers. Not the company's name you'll note, but the actual name of the man who built it ... and then carried it across the road to the testing centre, and then back again along with some dashboards and a brace of seats to the final assembly shop.

Aston's Newport Pagnell factory, where the Vantages are made, is completely out of date. It's a museum piece and loud like you wouldn't believe – a bit like the cars it makes, really.

I love going there, because it serves as a constant reminder that cars worked, and worked well, before robots and computers stuck their noses into the equation.

Indeed, it's probably fair to say that the Vantage is the last of a breed; the last of the heavyweight bruisers. One day it'll be gone and the world will be a little more boring as a result. **jc**

---

love all

Engine	V8, 5340cc
BHP	550
Brakes	And they work!
Top speed	185 mph
0–60mph	4.6 seconds
MPG	13
Optional	platinum AMEX
Today's cost	£190,677

The dash is fashioned from real wood and the leather really does look like it was once part of an animal

13

Guess what? You can't get an automatic version, and only the front windows are electric

Aston Martin Vantage

Audi quattro 20V

I'd driven the original

quattro on a number of occasions and considered it a nice little diversion, but nothing desperately special.

It revolutionised rallying of course, thanks to the grippy four-wheel drive system, but the heavy five-cylinder engine was slung way out in front of the front axle which made understeer a problem.

And there was turbo lag too. Like I said, it was an interesting car but I really didn't see it as a potential classic.

But then in the Autumn of 1989, I found myself driving what would turn out to be the last of the original quattros – the 20 valve.

And I can think of few occasions when a car made such a dramatic impression. Yes, the engine was still in completely the wrong place and there was an awful digital dashboard, but that car gripped and shifted like I simply could not believe.

The new 20 valve, 2.2-litre engine churned out 220 bhp and all of it was put to use though a new, torque sensing differential that sent power to whichever end of the car was best able to handle it.

Going into a corner it was as neutral as Betty Boothroyd

So, going into a corner it was as neutral as Betty Boothroyd. Then, coming out on the other side, power shot to the back where the weight was. Magnificent.

And safe. Coming up to a village, I saw the chance to overtake a lorry before we hit the 30 mph zone – so I dropped a couple of cogs and marvelling at how the turbo lag had been banished, surged past the truck …

And into the most frightening moment of my entire motoring career because there in front of the lorry was a car turning right – right across my bows.

There was nothing for it. I had to turn right too, even though I was doing … well let's say a little more than 60 mph. A right-angled bend when you're doing a mile a minute? Impossible?

Not if you have a 20 valve quattro, it isn't. That car, take it from me, is a life saver.

I guess it was a bit silly to have that hatchback rear end without fitting fold down rear seats – same as the Alfa GTV6 – but other than this, I maintained then and I still maintain now, that few cars have been so complete. **jc**

smartarse durch technik

593 YNV

I guess you think that I've gone a little mad here. Of all the thousands of cars that have been made in the last 35 years or so, I select an Audi saloon for inclusion in the Hot 100. In favour of a Ferrari 550, for God's sake.

Yes, and I'm a Jaguar fan too so what's this doing here?

Well though you might have seen no evidence – ever – I do have a pair of sensible trousers and I'm wearing them right now. And a cardigan with suede elbow patches. And a Vyella shirt.

The Jaguar XJR is the still best big saloon in my opinion but I fully understand why some people wouldn't go near it unless it was raining very heavily. They maybe had a Jag in the 70s and it broke down all the time. Or perhaps, they have a chauffeur and need more space in the back.

So that means they need to look at the offerings from Mercedes, Lexus, BMW and Audi. And in my opinion the best of the fifty grand autobahn stormers is the A8. And that's why it's here.

Don't be fooled by the all aluminium

body – it's not really much lighter than the competition, and anyway a car as good as this doesn't need gimmicks. It has four-wheel drive, and though I sometimes wonder whether it's worth it, I do think that if you're not charged a premium, it makes sense. And you aren't charged a premium here.

You aren't charged extra either for a wonderful-looking body which is topped off by the sexiest-looking wheels I've ever seen on a saloon car. I've seen a black A8 with blacked out windows tooling around Soho, and to my mind it is a staggeringly good looking car. Menacing to the point where you daren't step out in front of it, but graceful so that it seems to waft by in complete silence.

You can forget about the tiny engines – the best is the big 4.2 V8 which has so much get up and go, you begin to think it might be fuelled by coke. What matters here is the throttle sensitivity. The slightest twitch of your big toe sends the big saloon scuttling off like a small dog that's been dropped kicked by Rob Andrew.

With such urgency from under the bonnet and the monumental wet weather grip from its bang-up-to-date, four-wheel drive system, sport suspension is a tempting buy. But frankly, you're as well to go for the ordinary springs and dampers so the quality of the ride is preserved.

That way you have a car that does just about everything you could possibly expect. It's big. It's quiet. It's fast. It's fun. It's handsome and it's beautifully made.

The only trouble is that, as far as I'm concerned, it's not quite as quiet, or as fast or as fun or as handsome or as well made as the XJR. It is bigger on the inside though. So there you have it. The A8 is the perfect car for fatties. **jc**

double deutsch

Engine	V8, 4172cc
BHP	300
Brakes	ABS
Top speed	155 mph
0–60mph	7.3 seconds
Today's cost	£54,500

Bentley Turbo

Peer through the long list of Bentleys that have come and gone over the past 37 years, and a rather daunting thought begins to formulate. None of them should be in a book like this.

My grandfather had an endless succession of R types but while it was fun being picked up from school in these leviathans, I can't really say that they should be in a list of my 100 favourite cars. Especially as all of them made me feel carsick. And even then, of course, Bentleys were nothing more than rebadged Rolls Royces.

Now I didn't mind not having a Roller in these pages but I couldn't have slept at night if there wasn't a Bentley. But which one? The Continental T is fun but doesn't really justify inclusion. The Continental R

is a truly terrible ride and the ordinary saloons are Rollers in sports jackets. But then there is the Turbo which has no RR equivalent. It's a proper Bentley and having thumped one round Mallory Park last year, I decided it could come to the party, in about 99th place.

The basic problem is that the 6.75-litre engine is not all that quiet and there really isn't as much space in the back as you'd expect. And it doesn't handle anywhere near as well as cars costing less than half as much. And it depreciates like a Steinway falling off a tower block. The thing is though that it is pretty fast, and you should be in no doubt that it is fun to overtake great streams of slower-moving traffic in a Bentley. People are used to seeing them slithering up to

balls, rather then going balls out.

At first, you might think a car this big and this heavy couldn't possibly get round a corner at speed but by supplying it without a limited slip differential, all the power is simply spun away by the inside wheel. It's not much fun, but it is safe and it does startle old people.

And come on: doors which remain closed to people in a Jag or a Merc are opened to anyone in a Bentley. And that's fun too, especially if you bought yours second hand for as little as £30,000.**jc**

call me Mr Turbo

Engine	V8, 6750cc
BHP	385
Brakes	very, very big
Top speed	150 mph
0–60mph	6.1 seconds
Weight	Over two tonnes
Today's cost	£143, 268

BMW 3.0 CSL

A glam German with flared arches

This car came a long way before BMW was even half way to being a force on the world stage. They'd impressed us occasionally with stuff like the old 328 from the 1930s and the 1956 507, which provided so many styling cues for the current Z3. And there were a few raised eyebrows when they unleashed the wicked 2002 turbo. It may not have been the first turbocharged production car – the Chevrolet Corvair from 1963 holds that title – but it certainly showed Europe what a blower could achieve. It couldn't really be taken seriously though, because the chassis just wasn't up to that kind of power which leaped out of the engine bay without warning. On the press launch, nearly every single car was crashed.

No, the first truly serious BMW was, I think, the 3.0 litre CSL which exploded onto the motoring scene in the mid-seventies. BMW basically took one of their contemporary coupes and changed the boot lid, doors and bonnet from steel to aluminium, hence the L for lightweight in its name. To make sure everyone knew this was something special, a stripe was stuck to the flanks, and chrome wheel arch extensions were added – this was the 1970s remember. And because we are talking about the decade that taste forgot, the colour schemes were pretty wild too. There was red, white and a reasonable metallic blue but there was also orange and lime green. Inside, all was fairly normal except the standard car's comfy barge-like seats had been replaced with plastic buckets that didn't have any back rest adjustment. Under the bonnet, there was the usual 3003 cc, fuel-injected, straight six which fed its power through a standard four speed manual gearbox to the rear wheels which were held in place by standard semi-trailing arms and coil springs. So it was mechanically similar to the mainstream CSi, and it was no faster, but it cost more than £7,000, making it even more expensive than an Aston Martin DBS.

However, undaunted, the British importer of the time decided to buy 500 cars – half the total production. Industry observers thought him mad, and he did have to employ two people specifically to try and shift these preposterously expensive cars. But when the racing Batmobile version started to sweep all before it in saloon car racing, the bravado was vindicated.

I actually bought a CSL in the early eighties for £3,000 and it taught me more about driving than the BSM could have managed in a thousand years. You see, there was a lot of power, dodgy rear suspension and a lightweight body which meant that if it even looked like rain, the tail began to wag like it was connected to an over-excited dog. I could never fully relax because with no warning at all the rear end would snap out of line forcing you to dial in some opposite lock pdq.

It still holds a special place in my heart though, for two reasons. First, it's the only car I've ever sold for a profit and second, it was extraordinarily handsome. Somehow, the chrome wheel arches didn't look garish at all, and with no B pillar to spoil the roof line, it could turn the head of even the most dedicated motoring nitwit. There have been good-looking BMWs since but they've never quite managed to recapture the essence of the CSL.**jc**

BMW M5

Over the years, the BMW M5 has grown fat and complicated which is a shame because they got it right first time.

The original M5 was totally indistinguishable from a cooking model – and I mean totally. If you took the badge off, and many people did, there were absolutely no tell-tale signs that this 286 bhp monster could out drag, out pace and out corner virtually anything on the roads.

BMW's Motorsport division took a basic 5 series shell to its own factory where they added specially prepared suspension components and added their own blue printed engine – a version of the unit that had originally been seen in the M1 supercar. These were proper hand-built cars. And they felt like it.

They were designed and built by racing enthusiasts and it showed.

I took one to Cornwall once and loved it so much that on the way back to London, I went via Scotland.

Ordinarily, when I test a car, I borrow it from the manufacturers for a week, during which time I'll cover probably 500 miles. But I've just checked my old wall charts and it seems I 'tested' the M5 on eight separate occasions. I must have absolutely adored it.**jc**

BMW M1

The M1 showed what a muddle BMW was in, during the late 1970s. Even the name hints at confusion standing, as it does for 'mid-engined car – first type'... like they might need another try. It was built as a homologation special as BMW were determined to beat Porsche in Group 4 sports car racing. But not long into the design process, they gave up with the idea, and went into Formula One instead, making engines for Brabham.

The M1, meanwhile was in real trouble. It had been styled by Giugiaro while the car itself had been designed and developed by Lamborghini who were acting as sub-contractors. However, halfway through the process, Lambo went into one of its deeper financial crises and the M1 was therefore handed over to two other Italian coachbuilders – one doing the chassis, and the other, the plastic bodies. The two parts were married, hundreds of miles away by Baur in Germany. Amazingly 450 were made and it became quite a car. The 24-valve, six-cylinder engine was an early version of the M5's unit and gave the car a 162 mph top speed. Weirdly, it could do 50 in first. It handled well, looked good and was well equipped too. It even had a proper boot. But BMW's heart really wasn't in it and that's why it is so often overlooked in any trawl through the history books.

Not by me though. I'd put my hand in a bacon slicer if it meant I could own one.**jc**

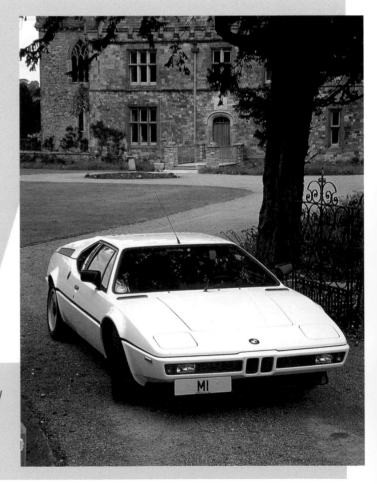

BMW Z1

An odd choice this, mainly because since the late 1980s, I've really rather hated anything with a BMW badge on the back.

The cars were fine, for those who wanted a quiet, Teutonic means of gobbling up miles in a vaguely sporty way but the people who drove them were a bunch of uncontrollable Onanists.

I have never been allowed out of a side turning by someone in a BMW and whenever you're on a motorway, heading toward a coned-off section, you can be assured that while everyone else will pull over in good time, there'll be a stream of Bee Ems, not getting into the correct lane until the last possible second. If this book were about the Hot 100 drivers, you lot in your 318s would be in the back, under the tosser section. Probably.

There is some doubt because my wife has a BMW – a margarine yellow Z1 with an interior to match. Tasteful it is not. But when it comes to waking up the pavement, no other sub-supercar even comes close.

Oh, the engine is a stock 2.5-litre straight six which means we're not talking about a desperately fast car. And even though it has a plastic body, it isn't particularly light or nimble either.

The rear suspension is complicated and clever – the Z1 was a rolling test bed for the 850 – but there are some corners around here that it really doesn't like. However, none of this matters because if you try to go quickly in a Z1, the wind will adopt the ferocity of a hurricane. And blow your ears off.

What you do if you have one of the 72 Z1s brought into Britain is drive round town slowly, letting everyone notice that it appears to have no doors. It does, of course, but unlike any other car they don't swing outwards or upwards; they slide downwards electrically into the sills. Which means that you can drive around with a dirty great hole in both sides of the car, leaning out if it takes your fancy, to light matches on the road.

This has been described as the greatest technological solution to a problem that doesn't exist but I don't care. Drop-down doors make the Z1 different. And a sports car should be different, and fun. It should be as good at raising eyebrows as it is at raising your heart beat.

Sadly, however, the Z1 was not what you might call a great success. Maybe the enormous £37,000 price tag put people off but I suspect the reason is more subtle. You see, it is impossible for any girl in a skirt to climb over the high, wide sills without flashing her knickers.

So, you may look cool while you're in it. And you will be cool if you drive it quickly. But if when you're climbing in you have to show everyone your crotch, you won't feel cool at all. In fact, under your arms and around your collar, you'll be hot, and maybe even a bit sweaty.**jc**

Citroën DS

Men with pipes and tweed caps were building cars out of wood in Britain, in the 1950s. At the same time, men named Chip and Rock were building cruise liners with wheels in Detroit. But in Paris, men with frogs legs on their breath were building a car so advanced that the Russians were rumoured to have tried abducting the chief engineer to work on their space program. The Citroen DS (which is French for Goddess, apparently) was designed and put into production in the mid 1950s. It had a revolutionary hydraulic suspension, enough room inside for most of the Dordogne and an amazingly rugged 1.9-litre engine. It also looked like Kermit the frog from the front, with its bug-eye headlamps and long, sloping bonnet.

By the late 1960s, however, the DS had gained a 2.1-litre engine, covered headlamps which followed the line of the steering as you turned corners, a semi-automatic gearbox not a million miles removed from the ones used in much later Formula 1 racing cars (although the DSs was plonked on top of the steering column), a leather interior as plush as anything in Whites and a reputation as the car of choice for President De Gaulle.

In the 1980s, the DS became the car of choice for men with too much hair gel who worked in advertising, and were keen on quoting French philosopher Barthes on the subject of their car. And there were still plenty of DS's left on the road, which was amazing given that 142 seperate different things could wrong with the suspension alone, and often did, all at once. The suspension fluid also powered the steering and brakes, and was carried in pipes which ran along the outside of the car's frame, thus being very susceptible to rust.

None of which mattered however, when you blew a tyre at midnight in the rain on a windy roadside. The car's pneumatic suspension had three different positions – two for driving (with a load of hay in the back, and without), and one for changing tyres. And it worked. You just had to remember to show the correct method of changing wheels to the local chain tyreshop to ensure they didn't put a trolleyjack through the bottom of your DS.

There's no denying that the DS looks beautiful, especially when moving. The seats are oddly reminiscent of those over-sized Draylon armchairs your Auntie Maureen bought second-hand in 1978, but they are damned comfortable, even when you corner with your foot

In the 1980s, the DS became the car of choice for men with too much hair gel

down and the back end drops lower than a snake doing the limbo dance.

A remarkable-looking convertible version was produced in the 1960s. The French have a phrase for it which translates as pretty ugly, let's just say that if the saloon was Jean-Paul Belmondo, the Decapotible was his younger sister. In the early 1970s, in a strange marriage, a Maserati engine was put into a DS coupe body, named the SM and priced at supercar level. The combination of DS hydraulics and Maserati engineering meant that few lasted long enough to get out of the showrooms. Any you might see today have undoubtedly been rebuilt.

A myth to clear up about the DS, is that it can't be clamped because it sits so low on its suspension when parked. It is not true. The back wheels are covered, and look impossible to clamp, but they are covered by panels which have to be unscrewed.

Finally, a question. Why, since when you start the car it has to raise itself to driving level, did the French police use them for so long? By the time a DS is ready to give chase, the crooks could be in Belgium.**jc**

I see Jeremy, so of all the thousands of cars which you could have put in your book, you choose to go for a jumped-up Jap coupe which has all the subtlety of a 70s disco and the manners of a drunk. Yup. Early 2.0 litre models with smooth flanks weren't much cop, I'll admit, but when the later 2.6 burst onto the scene with those ridiculous flared arches, I was in love.

It was in one of these that I managed to clock up my own personal best time for getting from the centre of Paris to Calais. And it was in a Starion that I enjoyed my longest-ever power slide – right across the golf course at Goodwood. I even took a Starion grass track racing late one night, which was about as much fun as I've ever had while sober. Well soberish, anyway.

I know it wasn't at the cutting edge of technology. It's just that whenever I had one on test, I always seemed to have a good time. So good in fact that I can even forgive its idiotic name. It was supposed to have been called a Stallion but the American importers, not realising that Japanese people often muddle up their ls and rs, got it wrong. So there you have it. A silly name, disco looks and a car that was nothing more than a Japanese Capri. But then again, you'll find the British Capri in this book as well. **jc**

Corvette Sting Ray

Little red

Engine	V8, 5359cc
BHP	250–435
Brakes	drums 1963–1964,
....	discs 1965–1967
Top speed	105–150 mph
0-60mph	5.4–8 seconds
How many?	21,513 in 1963
Today's cost	£18,000

A great empire is really only remembered for only one thing. The Greeks gave everyone democracy. The Romans gave everyone roads. The British gave everyone tea.

So how will the American empire be remembered a thousand years from now? The hot dog? Putting a man on the moon? Microsoft? Stealth planes? MTV? The atom bomb? There's no doubt that America has given the world more in the last 50 years than the world gave itself in the last 40 million. Which is why I'm always absolutely staggered to be reminded that a country which can invent and build a reliable re-usable spaceship, has never been able to make a sports car. Whoa. Here they come, galloping across the prairie, a million-strong army of Corvette enthusiasts, arguing that their beloved Chevrolet is a sports car in every way. To which I say, nonsense.

The most recently departed version spins so easily that if you park one outside a shop, it will be facing the other way when you come out again. All cars, even F1 racers, need suspension but that ghastly piece of plastic junk just didn't have any. I hated it. Foul is too small a word.

I hated the one before it even more. And the one before that was pretty vile too. But all of them pale into insignificance compared to the first example. This was a child molester in a sea of parking offenders.

However, there was one version which did come close to being vaguely sporty – the 1963 Sting Ray. With its plastic body it was light, and unlike any other American car, before or since, it was actually quite small too. Plus you could order one with a 327 cubic inch, 360 horsepower V8.

I drove one round Detroit for a day or

two and I must say that while it handled like a Hillman and rode like it had square wheels, it was a pretty amazing car.

Yes, the steering was nasty and it had drum brakes all round which meant it took longer to get from 60 to zero than it did to get there in the first place.

But I liked it nevertheless. Firstly, all Corvettes have looked wonderful, fantastic even, but this one was an aesthetic sensation. In hard top form with that split rear screen it was good, but as a convertible, it was almost perfect.

Either way, the '63 had louvres down either side of its bonnet and rotating

headlights – pointless and yet somehow essential – a bit like my electric pepper grinder which I mention only because it is my most treasured possession.

However, lots of cars have looked great. They need a little bit more than a nice face, and some louvres, to get in the Hot 100. And I've already said the Corvette is only vaguely sporty. So what's it doing here? Easy. With a big V8 under the bonnet, it makes the most wonderful throbby noise, encouraging you to double de-clutch even though it's entirely unnecessary. You hold onto gears a little longer than you should too, just to hear that big V8 bouncing its bass note off the skyscrapers. In this environment, you can forget, just for a moment, that you have no chance of stopping. Or that it's shaking your teeth out.

Let's be honest. We've all been out with dim girls who perhaps have a horrid laugh and questionable table manners. But you'll still take them to bed if they have the body of an angel.

My case rests. **jc**

> Let's be honest. We've all been out with dim girls who perhaps have a horrid laugh and questionable table manners

Datsun 240z

In 1969, Datsun was a new boy from a country that only 14 years earlier had been on the wrong end of two atom bombs. The world didn't like Japan very much and it wasn't that fond of its cars either. The only thing you could possibly say about the Datsun Cherry is that it had a radio. But then Japan surprised the world in a way that it had only achieved once before. Over Pearl Harbour.

This time though, America welcomed the invader with open arms. It was called the 240Z and the world had never seen its like before. It had a long bonnet, like an E-Type Jag, but under this one was an engine that didn't break down or piddle all over your garage floor. Inside, there were two seats but this was no spartan road racer like a big Healey or a TR. There was air conditioning for heaven's sake and, if you wanted it, automatic transmission. But that would have been a mistake because this was a real, genuine sports car. It looked like one and it went

like one too. Only it was priced to compete with dinner for two at a pizza joint.

You could have had one in 1969 for just $3,500 which was extraordinarily good value. And everyone knew Datsuns were reliable. And this one did 115 mph.

No wonder it sold in America at the phenomenal rate of 50,000 a year right up to the moment when Datsun decided to fit a larger engine, and ruin it. They put an even bigger engine in there and ruined it even more and then they gave up with the idea of making sports cars and concentrated instead on trying to make the world's nastiest car. Something they managed with the 300ZX. **jc**

not a Nissan

Engine.....	6 cylinder, 2393cc
BHP......................................	151
Brakes.............	discs/drums
Top speed..................	115 mph
0–60mph............	9.5 seconds
How many?.......	over 500,000
When?.....................	1969–1973
Today's cost ..	£5,000–£8,000

DeTomaso Pantera

Best colour: Primrose Yellow. Best Movie Moment: Cannonball Run.

Have you ever stopped

to wonder why Italian mid-engined supercars are so expensive? No, of course not; you've better things to do. But as I swan round my house in loose robes, I do ponder the issue from time-to-time and to be honest, I'm not really sure. I mean, they have the same amount of body panels as a Ford Fiesta, the same amount of wheels and both use the principle of internal combustion.

The problem, I think, has everything to do with volume. Churn out thousands of cars per week on a production line and your costs are low. Hand build a couple a month and they're sky high. So what if you could make a high-volume mid-engined supercar?

Well that's exactly what Argentinean race driver, Alejandro De Tomaso tried to do with the Pantera. He arranged for Ford to sell it through their dealerships,

saying it would do for Ford what the GT40 had done five years earlier – draw the crowds. Then he went back to his Italian factory and built a car that married Italian style with run-of-the-mill Ford engineering. Hey presto! A supercar for saloon car money.

But the volume just didn't come. Potential buyers knew it had a normal Ford V8 engine and felt that despite the 160 mph top speed, it just didn't belong. And those who did take the plunge found themselves with a car that overheated and then broke down a lot. Either that, or it broke down first and then they overheated. And the American versions, with pollution control equipment, could barely get past 140 mph.

Ford only managed to shift 5,000 Panteras before they lost patience and went back to making boxes. However, I'm still convinced that if you could find a

well-made De Tomaso, it's a damn good car with an engine that's easy to service, cheap to run and a doddle to tune. Plus you get a Giugiaro-designed body and a hint of Italian exotica. Later models sprouted ridiculous appendages like wings and fat arches, and with the volume gone, the price went ballistic too which made them nonsensical. If you're going to spend real money on a car, then you will buy a real car, like a Ferrari.

But those early versions which sold for £7,500 in the early 1970s made a deal of sense then, and look like a damn good secondhand buy now.**jc**

Dodge Viper

When I first tested the Viper, I remember climbing out and thinking that I would do anything up to and including robbing a bank to raise the fifty thousand pounds I'd need to buy one.

Chrysler, at the time, was trying to re-affirm its position as a force to be reckoned with, after years in the wilderness, dallying from time to time with bankruptcy. They felt that the time was right for a muscle car, cast in the mould of the fearsome Cobra and even the name – Viper – is a sort of homage to the car that inspired this thundering 8.0-litre beast.

Many say its V10 engine is lifted out of a truck but that's not quite true because in the lorry, it's iron whereas in the car, it's made from aluminium alloy. But don't think for one minute that this is a lightweight sports car, at home on a moorland B road. It's American, remember, which means it's heavy and very, very big.

I have driven one on some of the mountainous roads in the South of France and I learned that it's very nearly as wide as a Mini is long. A number of my colleagues, not realising this, found themselves going for gaps that were just too small. The damage bill must have been huge.

What I remember most about my time with a Viper on the Côte D' Azure though was the heat. Sitting in that cockpit, you are roasted not only by the Mediterranean sun but also by the massive engine and the superheated side exhausts.

Thankfully, air conditioning is provided

as standard these days, as is a roof. And that's a blessing too because in the early Vipers the buffeting, once you got past 80, would have made a North Sea life boatman turn for home.

The Viper is neither the fastest car in the world nor the best handling. It doesn't even get close on either front actually. And it is badly flawed in other departments too. But it's the car I used for leaving the church on my wedding day and that speaks volumes. You may only be a tiny speck in the vast expanse of its billowing bodywork but it makes you feel so damn special. **jc**

Subtlety, like irony, is not in American dictionaries. Which is why they build huge, great monsters of cars which can go extremely fast, and give them frighteningly accurate names.

The Dodge Charger is exactly that. With a massive 426 cubic inches (approximately 8 litres) of V8 creating 425BHP under a hood which stretches from Dallas to Houston, and a roofline which rakes at a perfect angle for practising ski jumps (and is long enough, too), the Charger can shift to 100mph in the time it takes to walk around it when it's stationary.

All of which is very fine indeed. And also very dangerous. The Charger can go very fast in a straight line. Since it was built for endless, straight, two-lane blacktops, where corners can take fifteen minutes to go around, not much time was spent designing the handling or suspension.

And since the Big Country is generally flat enough to spot Injuns approaching from a mile away, the brakes on the Charger were built accordingly.

Mind you, the boot is big enough to carry a very big parachute. **jc**

Ferrari Daytona

Given a free rein to decide on the cars that should go in this book, every single one would be a Ferrari. I'm like that. But you're probably not.

I had to be ruthless and for weeks pored over Ferrari books, Ferrari pictures and Ferrari brochures, trying to decide which models would not make the grade. It was a virtually impossible task but there was always one absolute constant – one car that would not be leaving the list – the 365 GTB/4. This is the car that took its nick name from the American city that bills itself as The Birthplace of Speed – the Daytona.

It is by no means the rarest Ferrari and though a few were raced, it doesn't even have a proper motor sports pedigree. It isn't the best-looking car ever to have emerged from the Maranello factory and nor is it anything like the fastest.

But despite all this, and for reasons, I can't really explain. It is my favourite Ferrari of them all.

The love affair began back in 1968 when my bedroom wall was a solid mass of car brochure centrefolds that had been 'stolen' from motor show stands. 'They are for my Dad, you understand,' I used to claim, 'he has an 1100 now but is thinking of maybe moving upmarket next year'. Obviously, pole position in this automotive gallery was the spot right next to my pillow, and that is where the picture of Ferrari's Daytona was glued.

The car was launched at the Paris Motor Show in 1968, amid gasps of surprise. The Lamborghini Miura had burst on to the market two years earlier with its engine mounted in the middle and everyone had thought Ferrari would do the same.

But they stuck with tradition and were even asking a premium for the privilege. At $20,000 dollars it was the most expensive Ferrari of all time and in 1968, the fastest too. One American journalist verified that it really would do 174 mph – a staggering achievement in a world that had not even been introduced to Sgt Pepper.

What made it all the more amazing is that we are not talking here about a light car. This was the full Big Mac and large fries with ice cream to follow. And two sugars in the coffee .

It was fast because its engine was quite simply a masterpiece. The 4.4-litre, double overhead cam V12 was fed by no fewer than six twin-choke carburettors so that it churned out a massive 352 bhp.

Small wonder it could haul itself over a quarter mile in 13.8 seconds, passing the finishing posts at a whopping 107.5 mph.

Americans had been used to European cars being light and nippy but for straight line speed, they reckoned we'd never be in the same league. Well the Daytona shut them up ... big time.

you are the one

Engine	dohc, V12, 4390cc
BHP	352
Brakes..........	discs all round
Top speed........	over 170 mph
0–60mph	5.9 seconds
How many?	around 1300
When?	1968–1974
Today's cost	**£70,000–£85,000**

Ferrari Daytona

And unlike their barges, the Daytona could handle too. It had an entirely new chassis which meant that despite the weight and the size – it was a BIG car – it could give almost anything a hard time on a twisty road. I knew all this, of course, in 1968 but it would be 25 years before I actually drove one. And I can remember the almost crushing disappointment as I nosed out into the traffic for the first time. The steering wheel was mounted almost horizontally, like in a bus which was apt because at slow speed, that's exactly what it felt like – a huge red Dennis. A lot has happened in the world of cars since the late sixties and I felt that no amount of nostalgia would disguise the fact that the Daytona was, to put it bluntly, old.

But then the road cleared up and I opened the taps on that Himalayan engine to be rewarded by a noise that sounded like God yawning. The V12 didn't sizzle, as you'd expect, it roared and began to catapult its charge toward the next millennium. But wait, here comes a corner, a nasty little left-right, zig zaggy thing that catches out even the most well-sorted modern cars. I'd better slow down ... a lot. No more than that.

Gratuitous Ferrari pit-lane shot

But there was no need because despite those tall tyres, the Daytona sailed through, belying its age with a lack of grip. The road holding, which is an entirely separate thing, was just sublime. This is a car that you can steer with the throttle, the brakes or the wheel depending on how you feel at any given moment.

For outright speed through a corner, a Peugeot 106 GTi would leave it for dead, but the Ferrari, by making you work at it, puts a far, far bigger smile on your face.

It's like going down to the garden centre and buying a rabbit hutch for your daughter's birthday, or getting out in the shed and making one. Which will bring the most pleasure?

Well the bought one will actually because if I'd had anything to do with the building process, the rabbit would escape and be found the next morning, on the doorstep, half eaten by a fox. But you know what I mean. You will also know what I mean when I tell you that this is one seriously good-looking car – in a manly sort of way. I've been trying to think of a humanoid equivalent and I think Sean Connery in his Untouchables phase is the best I can manage.

It's old, but the menacing, threatening presence has not be diminished by the passage of time – the price has though. You can buy a Daytona today for less than a 355 and while it's nowhere near as much fun to drive, it is more ... desirable.

And that, I suspect, is because of its place in Ferrari's history. It came at a time when the company was evolving from a race car team into a proper car maker. But more than that, it was the last of the front-engined models before mid-mounted motors became de rigeur.

Look at the 1990s 456 and 550 Ferraris and it doesn't take long to work out from which model their designers drew inspiration. It was the Daytona. It will always be the Daytona. **jc**

expensive – very

Engine.......	sohc, V12, 3286cc
BHP.....................................	305
Brakes..........	discs all round
Top speed..................	160mph
0–60mph............	6.5 seconds
Seats...................................	1.5
LM?..................	Le Mans, silly
When?.....................	1964–1965

When we think of Ferrari's racing operation today, we think of Formula One. But in the 1950s and 1960s, the greatest Italian sporting team of them all competed everywhere and at just about every level .

And they weren't just competing either – they were thrashing anyone who even dared to turn up and take them on. They won Le Mans nine times, The Mille Miglia eight times and the Targa Florio seven times. And, in the course of doing so between 1952 and 1972, took no fewer than 14 Manufacturer's titles.

Of course, in those days, race cars weren't specialist tools with slick tyres and engines that could only last 10 minutes – they were road cars that ordinary people could buy. I'm not kidding. You could actually pop down to your Ferrari dealership in 1965 and buy a 250 LM – exactly the same machine that had won that year's Le Mans race.

It wasn't exactly big on creature comforts in that the windows didn't wind down. And on the continent it must have been a nightmare because all LMs, to make them better on the track, were right-hand drive.

But it made a noise that could curdle soup from three counties away. Its 3.2-litre V12 engine was powerful too but more importantly, it was mounted in the middle of the car, behind the cockpit. There's nothing unusual in that these days – but in 1964, it was revolutionary. Dynamically it worked but they didn't really understand how to style them in those days. The front looked like it should house the engine and the back was a mess.

But who cares? This was a racer that you could take to the shops. And no-one in their right mind, could ask for more.**jc**

Ferrari 250GTO

At the height of the classic car price boom in the late 1980s, a Japanese businessman paid £10 million for a 250GTO, making it the most valuable car in the entire world.

Today, prices are down to the £3 million mark which means the man has taken a £7 million hit. But still, you won't find four wheels and a seat anywhere that costs so much.

It isn't even unique, but only 39 were made so it is rare. And it isn't really a road car either. I mean there's no carpet and some don't even have door linings. You can see the road screaming by through cracks in the body work. Like the 250LM, the GTO is a racing car that happens to be road legal.

I was once offered a drive in this remarkable car but I'm afraid I chickened. At £3 million, there is no way I would have dared do anything other than potter, and to potter in this is sacrilege. Like farting in a church.

So what then is the big deal? Well during its heyday, in the very early 1960s, the GTO swept all before it in endurance racing, even winning a hat trick of nine hour races in Kyalami. Because so few were built, pretty well all of them were competed at one time or another. This means that as you potter around today, you are sitting in the same seat as someone with nuts of steel who went out there in this glorious machine and fought, for hour after hour, to keep the V12

engine from tearing the skinny little tyres off the road.

In those days, motor racing was dangerous and sex was safe. And the GTO was the sexiest of all the racers. That's why its so valuable and that's why people will pay millions to own one.

And that is why you should look very hard at these pictures because one thing's for damn sure – you'll never see one on the road any more. Shame. **jc**

To potter in this is sacrilege. Like farting in a church

The other Ferraris I've chosen are here for very good reasons. All of them were, and still are, utterly fabulous to drive. But with Ferrari, towing poke and razor sharp handling – drivability in other words – has only ever been half the appeal. The other half is the styling. And that's why the 275GTS is here – because no Ferrari has ever looked better.

In the early 1960s, Ferrari was at a crossroads. Predominantly, it was still a race team but it was beginning to make cars that would never set foot anywhere near a race track. It reckoned it had an unassailable image and to capitalise on this, it had begun to make exquisite GT cars for the world's plutocrats.

To demonstrate the muddle this created, you only need look at the 275. The hard-top version had an all-new body and a 280 bhp engine because it was felt it would make a good competition car. Its soft-top sister, on the other hand, had a body based on an earlier car and a 260 bhp engine.

Now make no mistake, the coupe was a handsome brute, and it was fast and lovely to drive as well but it's the Spyder that I've chosen because you can just picture Brigitte Bardot careering round the south of France in one, her headscarf flapping in the wind. Or Isadora Duncan – her headscarf flapping around the wheel.

This is a car that should come as standard with a Michel Legrand sound track. The wire wheels, the gills in its front wings, the pronounced haunches – they were just so absolutely right. And to demonstrate that it was not really a racer, the GTS doesn't even look very good painted in red. For the best effect, set aside around £100,000 and get a blue one instead. And don't drive it anywhere, whatever you do. It may have been the first Ferrari to get a rear-mounted, five-speed gearbox to form a transaxle. Its all independent suspension was novel too, and its brakes had a nasty habit of going on strike from time to time.

The best course of action therefore is to park your 275 in your sitting room. You can tire very easily with most of the stuff on TV, except Top Gear perhaps, but you could never, ever tire of looking at this, the best-looking member of a startlingly handsome family. **jc**

call me Brigitte

Engine.......	sohc, V12, 3286cc
BHP....................................	260
Brakes..........	discs all round
Top speed	143mph
0–60mph .	around 7 seconds
How many?	200
When?	1964–1967
Today's cost	£80–120,000

Ferrari 355

Anyone who has driven a 355 will tell you the same thing. It is, without any doubt whatsoever, the greatest driver's car of all time. Oh sure, the world may be awash with more practical cars and it's short on nostalgia too, compared to other Ferraris like the GTO, say. But for hammering down your favourite bit of British back road, nothing on earth even gets close. Ferrari was in the doldrums throughout the 1970s and 1980s, which is why none of their cars from this period are in my book, but they came back with a vengeance in the 1990s.

I've always argued that to drive, modern cars are bound to be better than older ones. Anti-lock brakes made them easier to steer while stopping in a hurry. On-board computers made them more fuel efficient. Robots made them more reliable and new materials made them quieter.

There have been similar advances in suspension and tyre technology too so that they handle better. And of course, it is now possible to get more than 100 bhp per litre from an engine. The Daihatsu Charade GTti was the first production car to break this magical figure but as it only had a one litre engine in the first place, it didn't exactly worry Richard Noble. The 355 on the other hand has a thumping great 3.5-litre V8. And at 109 bhp per litre, that works out at a hell of a lot. There's no turbo either but each of the eight cylinders does have five valves for easier breathing – just one of the reasons why this incredible motor can rev to 8,500 rpm. If you forget to change gear, and you'll be tempted because when it's spinning this fast the noise is intoxicating, you can push it up beyond 9,000.

And at that point, it's screaming the sort of scream that makes you think a

It says to other road users that you are unburdened by the needs of mortal man

mine, all mine

Engine V8, 3496cc
BHP................................... 380
Brakes .... and ABS
Top speed .... 183 mph
0-60mph .... 4.7 seconds
Today's cost .... ????????

large dog has got stuck in the jaws of an even larger shark. And you are doing 155 mph. And then you are slotting the gear lever through it's open gate into sixth, heading for 185 or, if the wind is with you, a little more.

Obviously, engine technology is only half the battle though. Aerodynamics are important too which is why, if you look underneath a 355 you will find it has a completely flat floor. All the usual cables and pipes are hidden away under a panel that smoothes the air as it surges by.

The same sort of thing is going on up top as well but this is a damn sight more sightly. They do say that if something looks right, chances are it is right and if you want proof that this argument holds water, check out the styling of a 355. Not only does it scythe through the air with the efficiency of a scalpel but it looks like honey-covered heaven.

The 355 manages to blend both of Ferrari's greatest strengths in one car. It is superb to behold and amazingly, even better to drive. They've been at it now for 50 years, so obviously they must know a thing or two about fast cars. But the 355 shows that, in fact, they know more than the rest of the world put together. There is even space on the inside for two people, a mobile phone and in the front, under the bonnet, a hamster. That, though, is the point of a car like this. It's hedonistic and as a

Ferrari 355

result, it says to other road users that you are unburdened by the needs of mortal man. You have no children, just a hamster that you don't like very much.

I first drove a 355 soon after its launch and knew, after just a few miles, that I was experiencing motoring perfection – the automotive equivalent of a quail's egg dipped in celery salt, and then served in Julia Roberts' belly button.

In Italy, there was a sort of reverence from other people as I screamed by while in England, people are genuinely friendly when they see it coming. They pull over and let you past. Kids point, rather than spit which is what happens when you have a Porsche or a Rolls Royce.

I know all this because last year I did something unusual for a motoring journalist. I put my money where my mouth is and actually bought the car I was raving about. Yes, I bought a 355 and I never want to sell it. I know circumstances change and I may have to one day but believe me, it'll be the last thing out of the door, after the carpets, the washing machine and even, I've just decided, my bath.

There is something desperately relaxing after a day at the computer to get out there in a 355 and go mad, feeling its front end bite as you turn into the corner and then, as you pile on the power, the rear tyres fighting for grip. And all the time, you know that if they do let go, the resultant slide will be no harder to catch than a falling snowflake.

Better than destroying all the Nissan Sunnys in the world (almost)

You just dial in a little opposite lock and with tiny chirp from the tyres, and a little squeak from the hamster which doesn't like being thrown about like that, it's back on line, heading for the setting sun like it's a dog and the horizon's on heat.

Actually, I don't really keep a hamster in the boot but sometimes it sounds like it. This is because the carbon-fibre seats squeak against the firewall and the roof rubs against the windscreen frame. You'd notice these things in an everyday car. In traffic, they'd drive you mad. But I don't use my Ferrari every day. I only take it out when I fancy a drive and then, believe me, you wouldn't even notice if 14 entirely naked women came along and rubbed their breasts all over the windscreen.**jc**

Yes, I bought a 355 and I never want to sell it

Fiat 124

family coupe

Engine ...... 4 cylinder, 1608cc
BHP 110
Brakes .... discs all round
Top speed .... 112 mph
0–60mph .... 8.5 seconds
Today's cost £3,500

The Italians have made some of the most beautiful cars the world has seen, but they're generally more expensive than living for a year at the Monte Carlo Hilton with Naomi Campbell and Claudia Schiffer hitting the mini-bar in your room every night. Yet, being broadly a socialist country with a strong sense of family (i.e. non-contraceptive practising Catholics), Italian car manufacturers understand the importance of having good-looking cars which can comfortably seat a family, and also go very fast. The Fiat 124 series of the 1960s was the embodiment of the go-faster family ethic. The saloon was a little boxy, but despite looking small from the outside, had more leg room than First Class on Virgin airlines. The Claudia Cardinale of the series though, was the 124 coupe 1600. A sleek, twin-headlamped sports car, it had a boot that could carry luggage, room in the back for two girls in miniskirts, headroom for even a very tall Italian, and the power to move to 60 from standing in around 9 seconds, and top 112mph. It was also surpisingly rugged for a late 60s Fiat.

It was the stuff that urbane, would-be sophisticate young family men (and they start early in Italy) yearned for. It had the look of a car with its radio permanently tuned to Henry Mancini, and its steering pointed toward winding mountain passes.

There's not been a better production-line 2+2 sports car to come out of Italy since. (Hot hatches don't count).**jc**

Fiat Abarth 695

Imagine Noddy travelling so fast that Big Ears gets his protrudences flattened without recourse to the surgeon's knife. That'll be the Fiat 695 Abarth. Based on the Topolino 500, racing experts Abarth upgraded and bored out the original engine so that instead of poodling around at the speed of a snail with TB, the little car which was designed for dwarf Italian families with seats as comfortable as the electric chair, could hurtle across Milanese pavements at the speed of a snail with a rocket backpack.

Mad Italians subsequently raced and rallied the ridiculous mini with some success. It deserves inclusion for the fact that it is possibly the most bonkers motor made outside of the Far East. Why would Noddy want to go rallying?**jc**

Ford Cortina 1600E

Quick. Think up a list of British institutions. Chances are you'll have the Queen at number one, with drizzle and Stonehenge vying for the number two slot. Until I draw your attention to the Ford Cortina.

Ford had already enjoyed spectacular success with the Mark One but it was the Mark Two that put this remarkable car on the map. With its glamorous name, its boxy but handsome looks and its low price tag, it was a seminal moment in car marketing. In the rapidly expanding fleet market, Ford's advertising boys told us it was THE car to have and with just about every sales rep plying their wares from the boot of a Cortina – my Dad included – it went to number one in the sales charts. And simply refused to budge.

But what about the 1600E? Well when my father stopped flogging timber to education authorities and began his own business, the Cortina was a casualty and we ended up with a dreadful Austin 1100 instead. These were dark days. But when business picked up there was only one car which properly advertised his success – the Cortina of course, but this time, the 1600E. Under the bonnet, there was a perfectly ordinary 1600cc cross-flow engine which drove the rear wheels through a four-speed gearbox. It could hit 110 mph which was pretty meaty in 1969 and it could corner with gusto too. But this was not really the point. No, the 1600E's major trump card was its trinketry. It sat on Rostyle wheels which were just the coolest shoes any car had ever worn. At the front, there were extra spot lamps and at the back, a black panel beneath the boot lip. It was lowered too, for a sporty stance but it was inside where Ford had gone really mad. In the centre of the dash were four chrome-ringed dials but people didn't really spot these because the dash itself was real wood, as were the door cappings. This was a rep-mobile in a frilly dinner shirt.

When my Dad picked me up from school in his brand new 1600E, I was almost beside myself with excitement and pride. Everyone else had a Cortina but we had a 1600E. So, when I passed my driving test, there was only one car to have. It cost me £800 and it was the flashiest Ford of them all ...

I resprayed it, used paint stripper on the wheels to give them an all chrome look, added even more lights to the front and glued fur to the door linings. I even had a picture of Debbie Harry in the centre of the leather steering wheel with its drilled and polished aluminium spokes.

It went wrong quite a lot but it was the only car I've ever owned that I could usually fix. You just found the offending part and hit it with a hammer, and then it would spring into life. Try that with a modern engine management system and see how far you get.

I know the 1600E has no place in anyone else's Hot 100 but your first car is like your first girlfriend. You never really forget her do you? **jc**

41

Ford GT40

Younger readers beware. When asked to eat your greens so you grow up to be 'big and strong', go directly to your room and refuse to come out until your parents have died of old age. If they fight back by breaking the door down and force feeding you with spinach and broccoli, ensure that afterwards, you down a pint of salt water. Call Princess Diana and find out how to catch Bulimia. Greens, take it from me, have a nasty habit of making your adult life a complete and utter misery.

Ever since I was six, I have had a burning desire to drive a Ford GT40 but because I was made to eat cabbage until it was coming out of my ears, I am now too tall. By about nine inches.

I've always argued that I can get into any car if I really want to. I'll put up with extraordinary discomfort in a Ferrari F40 or a Jaguar XJ220, neither of which are really able to take someone who's 6' 5". But for me, the GT40 was a car too far.

I'd been looking forward for months to the day when I would film Ford's first supercar. I'd written a script. The crew was booked. Ford had agreed to lend me their prized museum piece and we had a

country house hotel to use as a backdrop. Top Gear's new producer even came down to see how the filming process worked.

But it all went horribly wrong because my legs simply didn't fit under the dashboard which meant I couldn't settle down in the seat, and that in turn meant I couldn't close the door. Or reach the pedals. For an hour, I tried all sorts of different entry procedures but short of ripping out the dash, or amputating my knees, it wasn't going to happen. I was never going to drive a GT40.

Some have subsequently argued that one GT40 was built with lumps in the roof to accommodate Dan Guerney, the vertically plentiful American racing driver, but headroom isn't the problem. It's my legs that will always keep me and my dream car apart.

In some ways, it's not the end of the world because old cars are usually long on nostalgia but short on driver appeal. I would probably have hated it. But then again, I probably wouldn't. And now, I'll never know.

I do know this though. The greatest thing about the GT40 is not how it looks, though it is a wonderful piece of design.

For four glorious years,
Ferrari could do nothing
about a Ford that
had been built out
of pure spite

Ford GT40

Nor is it the way it goes, though it can top 200 mph in race trim.

No, the best bit is why it was built. In the early 1960s, Enzo Ferrari was ready to sell his company and Henry Ford was keen to buy. Negotiations were going well until Enzo decided his baby would be suffocated by the weight of Ford bureaucracy and pulled out.

Henry was livid and went back to America, demanding that his engineers build a car that would thrash Ferrari on the race track.

The American arm of his vast operation began work on the V8 engine while the British division set to work on the chassis and body ... and the GT40, so named because it's just 40 inches tall, was born.

It had an inauspicious start in endurance racing, which must have made Enzo a happy man. But in 1966, it wiped the smile off his face by winning at Le Mans, something it did again in 1967, 1968 and 1969. For four glorious years, Ferrari could do nothing about a Ford that had

been built out of pure spite.

Having proved he could meet and beat Ferrari on the race track, Ford added a boot to the GT40 and tried to sell it as a road car but poor press reviews caused him to pull the plug.

And Ford has never built a supercar since. Yes, there was the ill-fated GT70, the disastrous RS200 and the GT90, which never even got past the concept stage, but I suspect they know they'll never top the glory days of their greatest car – the GT40. **jc**

Ford Mustang

When you bought a car before the Mustang came along, there was no such thing as an options list.

In the very early days of car making, you bought a chassis and had a coachbuilder design a body to suit your personal requirements, but all that stopped when mass production hit the scene.

To keep costs down, every car rolling down a production line was exactly the same as all the others rolling down the production line. This is what prompted Henry Ford to say that 'you can have any colour you like so long as it's black'.

However, the Mustang changed all that. Ford decided that their new car would be available with a choice of engines, a choice of body styles and most amazing of all, a choice of interior fixtures and fittings. This meant you could have a basic Mustang saloon for less than $3,000 dollars but, if you felt so inclined, you could go bonkers and pay double that for a hot V8 convertible with bucket seats. It was a risk, and Ford were cagey, saying that they expected just 100,000 sales in its first year. But as the 12-month period finished, they had sold a staggering 681,000 Mustangs, making it the fastest-selling car of all time; a record that has never been beaten.

It wasn't just the options list that had won the customers either. It was a damn good car that could be bought for not much money, and tuned for pennies.

This is probably why Hertz even offered ultra-fast Shelby Mustangs for hire. Until they realised that people were renting them for the weekend, and entering them in race meetings. And remember, these were basically the same cars that were being used for the school run all over the American suburbs.

Since the heady days of the mid sixties, there have been countless Mustangs – some good, and some, like the post-oil crisis model, absolutely awful – but Ford has never really been able to recapture the magic of a '65 V8 drop top.**jc**

It was a fast Cortina in a party frock

America had the Mustang and we had a Cortina with a rakish body. It wasn't fair right up to the moment when Ford slotted their Essex-made, 3.0-litre V6 under the bonnet.

Yes, it had exactly the same suspension as a horse and cart which helps explain why Bodie spent his entire time in CI5 skidding though market stalls. But it was fast in an honest to goodness sort of way. Think of it as food. If you're hungry and just want something to eat, there's no point cooking braised duck in a ginger and cherry sauce. Such a dish has a place – in the bin if I'd cooked it – but only on special occasions.

It's too much of a bother when you're tired and have just got back from work at 10pm. That's when the fish and chip shop comes into its own.

And so it goes with the Capri. It could get you home in the same way that fish and chips fill you up. And that easy to manage and easy to maintain 3.0-litre engine could make it exciting in the same way as salt and vinegar can very cheaply be employed to kick start a potato.

Eventually, Ford began to suffer from delusions of grandeur though and fitted a fuel- injected, 2.8-litre German engine in their Capri, along with leather seats, thinking that it was a serious coupe.

They got it into their heads that it was Dover sole with French fries. But it wasn't. It was a fast Cortina in a party frock. And you can't buy cars like this any more in the same way that there are now more curry restaurants in Britain than fish and chip shops.

We're all after the exotic these days and it's a damn shame.**jc**

Ford Transit

In 1996 Ford ran the 3 millionth Ford Transit off their production line. It was immediately fitted with Wolfrace wheels, a furry dashboard, the engine replaced with a V4 Corsair 2-litre engine and a mattress thrown in the back.

Where would Slade, the Sex Pistols, the Smiths, Oasis and these days Blur, be without a Transit? They'd be sitting on the M6 in an overheated Bedford van trying to thumb a lift to their big break, that's where.

The Transit has been the making of every British band since the first one rolled out of Dagenham. It's also been the making of countless teenage Lotharios since it was discovered that you could get a six-foot mattress, mega-watt hi-fi, boat-shaped bar complete with spirit dispensers, fake fur and a spinning mirror-

ball into the back, and still have room to swing 40" pantaloon flares. Second-hand Trannys were cheap, rugged and ripe for customising. Side-pipes, go-faster stripes (with obligatory filler over the flared arches) and fur-trimmed whiplash aerials were de rigeur for the wannabe David Essex of Ongar.

Huge padlocks, blocked-off rear windows, wilting suspension and concrete lagging were essential for builders.

Plain white, a Ford Zephyr V6 engine and quick-release rear doors made it a villains' stagecoach. Wire-meshed windows, funny numbers on the roof and a people-crusher front bumper made the Tranny the SPG vehicle of choice.

Oh yes. You can take a Transit out of Essex, but you can never take the Essex out of a Transit.**jc**

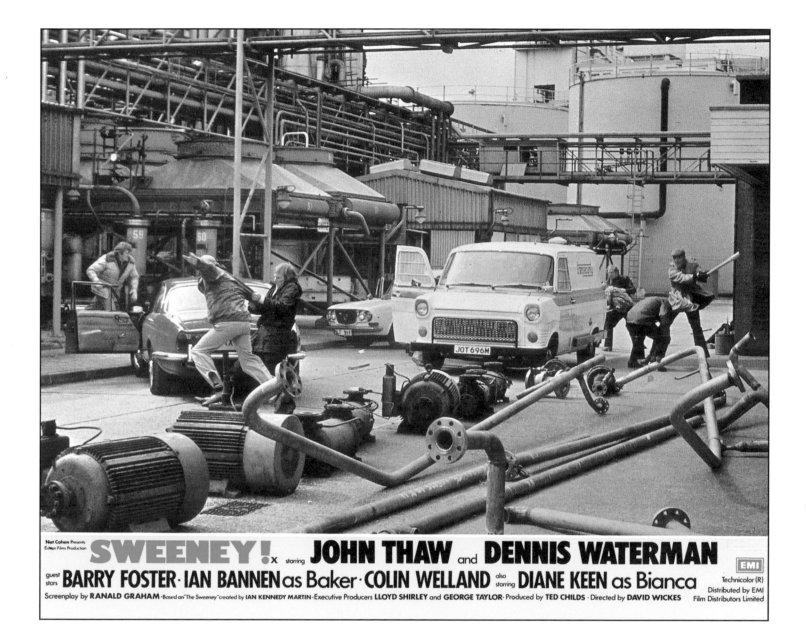

SWEENEY! x starring **JOHN THAW** and **DENNIS WATERMAN**

Nat Cohen Presents
EMI on Films Production

guest stars **BARRY FOSTER · IAN BANNEN** as Baker · **COLIN WELLAND** also starring **DIANE KEEN** as Bianca

EMI
Technicolor (R)
Distributed by EMI

Screenplay by **RANALD GRAHAM** · Based on "The Sweeney" created by **IAN KENNEDY MARTIN** · Executive Producers **LLOYD SHIRLEY** and **GEORGE TAYLOR** · Produced by **TED CHILDS** · Directed by **DAVID WICKES**   Film Distributors Limited

A Ford Zephyr V6 engine and quick-release rear doors made it a villains' stagecoach

Ford Escort RS2000

When you first hear of the 'Young Farmers' you might think that these people are in fact farmers who are young. But they're not. Young farmers are in fact a group of people who have little or nothing to do with farmland except they crash into it a lot.

Young farmers, as far as I can work out, are young country people who go to the pub at night, drink hundreds of pints of beer and then drive home very quickly in cars bought for them by their parents – people who are farmers and who

therefore have massive EU subsidy cheques to spend on their offspring.

When I was young and lived in the countryside, all the best pubs were full of young farmers – you could spot them a mile off in their Vyella shirts and sensible shoes. But the trouble with sensible shoes is that in order to make them robust, they have to be large. And large shoes have a habit of hitting two pedals at the same time on the way home, especially when they're attached to a leg that is being fuelled by a mixture that is two-parts blood to 98-parts beer. What made life even more dangerous for oncoming traffic is that all young farmers had Ford Escort RS2000s which were connected to the road by a suspension system that shared much in common

with the foundations of Westminster Abbey. Forgiving is not perhaps the best word you could choose to describe it, and that was a nightmare because the engine was actually quite powerful. An RS2000 could get you into trouble before your addled brain knew what was going on. And then, no matter what you did with the brakes, the throttle or the wheel, the suspension would not get you out again. The RS2000 became so adept at bouncing through fields, that it became the de rigeur rally car of the day, sweeping all comers aside as it ricocheted from tree to tree.

It was tough, fast and wild and I must confess, I thought it was the bees' knees. Still do actually, but you can keep that quiet.**jc**

The de rigeur rally car of the day

Ford Escort RS200

If you give motor sport engineers a free rein, they'll build cars that no driver could possibly handle. That's why the Formula One rule book is as thick as school pastry and that's why, in rallying, Group B cars were banned.

It all started with the quattro which brought four-wheel drive and turbocharging to the forests and, pretty soon, everyone was at it. The rules said that 200 examples of a car must be built before it could be entered on the world stage, but 200 was nothing, which meant the engineers went mad. Ultra-short wheel base, mid-engined, four-wheel drive rockets introduced rally drivers to the kind of speed hitherto only experienced by USAF jet fighter pilots. There was the Lancia Delta S4 of course and the Peugeot 205 T16, then Rover got in on the act with the Metro 6R4. In a desperate bid to keep up, Audi made a limited run of short wheelbase quattros but the engine was still in the front, which meant there was still understeer, and therefore it was never in with much of a chance. All this was in full swing by the time Ford woke up and decided they wanted to play as well. They began work on what was to become known as the RS200, and asked Reliant to build the bodies.

Not an auspicious beginning then but the end was even worse. Soon after it came out, an RS crashed in the Portuguese rally, killing spectators. Rallying's governing body decided enough was enough. Group B cars were banned and the homologation rules changed to say that in future, 5,000 cars were to be made before a car could be competed. The RS200 was superseded therefore by cars like the Sierra

Cosworth before it really had a chance to shine – or sell. Ford were left with a warehouse full of the things, and on a cold bleak February I took one for a spin round the company's Boreham test track. It was brutally fast and rough as a badger but it looked like an angel.

If it had won a championship of two, the 200 examples would now be worth a fortune. As it is, they're smashed to pieces every weekend like so much junk on the world rallycross circuit. Pity.**jc**

Ford Escort Cosworth

Insurance group 20 — the same as a Ferrari!!

Funny isn't it. There are five Ferraris in this book and eight Jaguars but the greatest number of entries from a single manufacturer is Ford. This is because from time to time, the American giant gets bored with making run-of-the-mill rep-mobiles and goes completely bonkers.

I think it's fair to say, that their crowning achievement was the Escort Cosworth. It couldn't have come along at a better time. The Escort on which it was allegedly based, was a dog. It suffered from body flex, woeful reliability problems, and even if you got a good one, it was truly terrible to drive.

Ford urgently needed a boost for this dog and got one, big time, in the shape of the Cossie. There'd been a Sierra Cosworth before, of course, and the hot 500 version followed by the Sapphire 4X4 was a serious piece of kit, on the race tracks, on the rally circuits and even on the open road. However, the Sierra had

been replaced by the Mondeo, so Ford simply fitted a modified Escort body to the Sierra's undersides and the greatest Cosworth of them all was born.

So what about that wing then? Well calculations have shown that it took about 12 mph off the top speed but couple its effects to the deep front air dam and you had the first production car ever to have negative lift. At speed it was literally being pressed into the road, making it harder to unstick. However,

none of this really mattered because that rear wing's main job was to make people sit up and take note. It was there to make people who'd bought the horrible cooking Escort feel a little less annoyed about their decision.

And the same could be said of the big, fat wheels, and bulging wheel arches. This was a damn clever styling exercise because it had all the garnish needed for rallying homologation purposes but also looked wonderful on the roads.

And it was an Escort, an everyday car for everyman with a realistic price tag AND a very real ability to rip up the Tarmac and run rings round Percy Porsche. I ran a Cossie for a couple of years and everywhere it went people pointed and shrieked. In Essex, where Basildon man could make or break a government, this was THE car.

So it's a shame really it was built in Germany and killed by a bunch of legislature-mad twats in Belgium. **jc**

fat boy

Engine..16 valve, 4 cyl, 1993cc	
BHP	227
Brakes	discs all round
Top speed	138 mph
0–60mph	5.7 seconds
Really?	Yup
Today's cost	£11,000

Optional extras overdrive, automatic gearbox, stick-on red dragon

Gilbern Invader

The Welsh have to put up with a lot of abuse from their neighbours. Beaten at rugby, football, hockey and probably even tiddlywinks, they are the butt of jokes about sheep and lamp posts from Lands End to John O'Groats. In any competition between the four Home Nations, Wales invariably come in fourth.

Except, that is, in terms of producing cars.

Name me one proper car built and designed in Scotland or either of the Irelands. The DeLorean doesn't count, having been paid for by Colombians. Can't think of any, eh? Now name me a car built, designed and produced in Wales – and there you have it, the Gilbern Invader. It's plastic, came in kit form so that men with beards and woolly jumpers could spend five years putting it together, had that versatile 3- litre V6 Ford engine under the bonnet and enough room inside for two grown men and a sheep on the back seat. The Invader sported snazzy wheels, a big red plastic dragon on the front, came in gold, green, white or red and could cruise in comfort at over 100mph, although you'd be advised to only attempt that in a factory-built example. It was in effect the saloon version of Reliant's Scimitar, but didn't suffer the stigma of being the least-liked Royal's favourite mode of transport. So it's here because it's not a bad car, but most importantly it's Welsh. **jc**

hello boyos

Engine	V6, 2994cc
BHP	144
Brakes	discs/drums
Top speed	129 mph
0–60mph	9 seconds
Body	plastic
Colour	white or gold
Today's cost	£5,000

Honda CRX

Someone married me because of it and then, six months later, left and took the damn thing with her

I would never ordinarily go out there and actually buy a Japanese car. Not unless I'd become mad in some way.

But in 1983, I saw a tiny, tiny picture of an even tinier car that had been launched at the Tokyo motor show and I knew I had to have one. It was called the Honda CRX and I told myself that it would be ideal in London, fast enough to get away from the lights promptly and yet small enough to fit in motorcycle parking bays. I knew it would be reliable too because it was a Honda.

And it was. In five years, I never washed it, or serviced it but even when there was moss growing up the back window, it never even thought about failing an MOT. But this wasn't really why I bought it, you know.

Years of Scirocco ownership told me that small, sporty cars like this have an appeal among women and that I'd be halfway there once they saw how I'd arrived. It worked like a dream too; so well, in fact, that someone married me because of it and then, six months later, left and took the damn thing with her.**jc**

Honda S800

For the last ten years, Japanese motor manufacturers have been trying their hardest to recapture the essence of classic Sixties British sports cars. They even have huge computers working out the exact tonal pitch of the exhaust fitted to a non-rusted MGB (although that they got one as source material is hard to believe). Apparently, to young Japanese professionals, the lure of wind and flies in your hair, seats so low that you can see under passing low-floor buses, and an engine not powerful enough to disturb Lionel Blairs hairdo, is irresistible.

So why then, have Honda not resurrected this, possibly the prettiest two-seater sportscar to come out of Japan in the Sixties? There were two versions, one powered by an awesome 598cc engine, the other by a mighty V4 791cc engine pushing out 70 horse power at 8000 revs, which Honda claimed could reach a top speed of 99mph, and get to 50mph in 8.5 seconds. In 1969 the S800 cost £725, which was £100 more than the MG Midget – but then the Honda is a good deal faster and better looking. I know which I'd rather squeeze into.**jc**

safe and speedy

Engine	4 cyl, 2157cc
BHP	182
Top speed	138 mph
0–60mph	7.3 seconds
MPG....................................	31.4
Today's cost	£23,000

Honda Prelude

Unlike all the other Japanese car manufacturers which are large corporations designed to make money, Honda was started by one man who wanted to make cars. As a result, his machines have always had a bit of flair and passion while Toyotas and Nissans have tended to be ultra-reliable boxes.

The Prelude, in many ways, is a typical Japanese car but I would always put it in a list of all-time fave raves because its V-TEC engine is just so amazingly sweet. At ordinary speeds, the 2.3-litre motor is quiet and unobtrusive but take it above 4,500 rpm and the profile of the camshaft changes, giving it a sporty edge. As you get up toward 8,000 rpm, it's screaming to the point that you really don't want to have to change gear. But you do and then you're treated to the adrenaline- pumping rise in noise all over again.

I've said before that this is the best mass-produced engine of all time and I'll say it again now.

It doesn't really matter which car you put it in, but I guess that the 1995 Prelude was perhaps Honda's best looker, so that's the one I've picked.**jc**

Jaguar XK150

At the end of the 1950s, Jaguar had a huge racing reputation which they had miserably failed to translate into road-going potential. The futuristic-looking C-Type and Batman-like D-Type were winning races by simply turning up. But on the road their sporting coupes were looking and feeling a little dated. At the time, of course, the Jag men in brown coats were busy distilling the essence of the E-type, but in the meantime they had to do something. So the XK150 was born.

Based on the old XK120 and updating the 140, the comfier, heftier 150 used much of the old technology, but married it to disc brakes all-round which had been successfully tested on the D-type. By the beginning of the Sixties, the 150 had found itself tarted up with three massive carbs that resembled Thunderbird wine bottles and a bored-out block which carried what looked like Mother's horseless carriage to 60mph in seven seconds, and topped 135.

Which is why it's here. This is a car that

looks the essence of English gentility and smacks of a time when men wore caps, women wore headscarves and children were presented for inspection before bed by Nanny. Yet it can perform like a modern-day monster, has a lovely, throaty-sounding exhaust and doesn't look daft with whitewall tyres.

The E-type was undoubtedly ground-breaking and unique, but alongside it's immediate predecessor, it also looked almost obscene.**jc**

Jaguar E-Type

I was always well aware that Jaguar was a fearsome racing team in the 1950s and that its C-Types were cars to be dribbled over in the pits. I also knew the XK150 was a gentleman's performance car. What nobody knew until 1961 was that Jaguar's new designer Malcolm Sayer was a bit of a kinky guy. Ask anybody about the E-Type and the first thing they'll mention are the words phallic, and symbol. Where previously Jaguar had used loping curves and feminine angles, here they unveiled a decidedly masculine motor. The fact that it was very cheap – about the price of a bus ride along Oxford Street – and could manage to reach 150mph made it the early 1960s motoring yob's car of choice.

Although originally only a 3.8 update of the XK series under its frightening bodywork, by 1965 and the introduction of the 4.2-litre straight-six cylinder engine, the E-Type had become the subject of many a grown man's wet dream. This, in its roadster styling, is the ultimate E-Type, the so-called series 1.5. Later V12 cars may be a little more powerful, but they don't look as good as the earlier cars. Anybody growing up in the 1960s, even people not remotely interested in cars, could not fail to be a little bit in love with the E-Type. It was the Beatles, Stones, Herman's Hermits and Dave Clark Five all rolled into one. Irresistible.**jc**

sex on wheels

Engine	dohc, six, 4235cc
BHP	265
Brakes..........	discs all round
Top speed	150 mph
0–60mph	7.5 seconds
MPG	20
When?	1965–1967
Looks	Sex on wheels
Today's cost......	£12–£30,000

Jaguar MkX

A car that demands respect in an old-fashioned, East End kind of way. The makers of The Krays film gave Reggie a Mark X to drive. Many an old lag, reminiscing about the good old days when the bruvvers only smacked their own kind and made the streets safe for normal people, will tell you that the Ten was the car of choice for top blaggers. And it wasn't just because it was big enough to get four Arfur Mullards in the back, and had the XK150 S-type 3.8 under its never-ending bonnet.

No, the crooks loved the Ten because it looked the business. It's big, frightening, expensive and almost American enough to carry George Raft and Jimmy Cagney to knock over banks in Chicago. It was the underbosses' chosen mode of chauffeured transport (the big bosses going for Merc 600 Pullmans). On special jobs, it would be used as the getaway car. The Ten could also take any amount of ramming, so wages' vans were literally easy to knock over. The police, many of whom wanted to be the blaggers but preferred the steady pay and uniform, soon cottoned onto the Ten's appeal, of course. Instead of the humdrum Wosleys and Rileys they had to poodle around in, they started the car of choice for crooks, which they could never catch once it reached long, straight roads. At which point it was all over for the Ten. Crooks wanted a car that the police couldn't afford. The police inevitably wanted whatever the crooks had.**jc**

Big enough to get four Arfur Mullards in the back

Jaguar MkII

If you've never seen Michael Caine in a British film from 1971 titled Get Carter, then I suggest you do. Caine is a nasty piece of work, a hitman whose brother is killed in Newcastle by small-time crooks, and who sets off to avenge him. He drives to Newcastle in a 3.4 Jaguar Mk II, and in the process gets the closest any British film has ever managed to making the A1 and surrounding roads as romantic as America looks in all those Hollywood road movies.

The MkII has always been the British filmmaker's preferred motor. Since the 1960s a whole bunch of gangster films have featured one or more – Robbery, Villain, Performance, Callan, Mona Lisa.

That's because it looks great on the move, with light reflecting from wet roads onto the gleaming bodywork, the single headlamps and leaping statue tearing through bleak landscapes. The Mk II made a perfect visual irony, the car of respectable upper management subverted by dirty wrongdoers. Or some such rubbish.

Despite all that, the Mk II 3.8S is a damned fine car. Comfortable, fast, good-looking, it could get to 60mph in around 8.5 seconds, and there was room in the back for an office outing.

I'd drive one, even despite the fact that it's the same car that grumpy old Inspector Morse uses to float around Oxford.**jc**

I think it's a fairly well-known fact that I'm a huge fan of Jaguar but this has nothing to do with Le Mans or the E-Type. My enthusiasm for cars was nurtured during the 1970s, when Jaguar was just an item on the news about industrial unrest and strikes. And the few cars that did get out of the factory gates, never got further than Birmingham before breaking down. In those days, I was far more interested in the emerging power of BMW whose new 7 Series seemed to be everything the Jaguar XJ was not: like it worked and the Jag didn't.

But I remember precisely the moment when my mind was changed, forever. I was in a Series Three XJ12, leaving the Chiswick roundabout and heading up the slip road onto the elevated section of the A4, and suddenly the meaning of automotive life became clear.

Until then, I'd always assumed that when a car accelerated, there was a lot of noise, signalling to the driver that the engine was being asked to work for a living. All cars did this. But the Jaguar did not. I pressed the accelerator right to the floor and for an instant, actually

believed that something had broken. This was a strong possibility, remember, because this was a Jaguar from the bad old days. Maybe the throttle cable had snapped or maybe there had been some catastrophic engine management failure. Either way, I had buried my cowboy boot in the Wilton, and it was as though I'd put my foot in a bucket of treacle. There was complete silence. But then I noticed that the bonnet of the car was pointing at the sky and that the speedo was shooting round the clock as though it were trying to escape from a burning building. In less

Jaguar XJ12

It was like a cottage in there with standard lamps and a wood burning stove for a heater

than eight seconds and in almost complete silence, the V12-propelled XJ was hammering along at 60 mph, but there was no sign of any let up. The three-speed gearbox blurred the changes and the eerie all-enveloping quiet persisted all the way up to a speed that would have put me in court and a lesser car in the Armco. I was staggered.

I don't ordinarily lift the bonnet of a car but when I got home that night, I lifted the flap on the Jag and was genuinely surprised to see something that used internal combustion under there. I'd expected a white panelled room with a dilythium crystal and Scottie running around, urging ever more from the warp core. It wasn't even a handsome engine but it had won me over, big time.

Now I know the Jag V12 is out of date these days and that it costs a fortune to service and feed but when it comes to isolating plutocrats from noise and stress, no other motor on earth even gets close. And let's look at the car in which it was mounted – the Series Three. What a staggeringly lovely car that is. They'd got the grille just right by then and the pepperpot alloys were a masterstroke of restraint with a hint of sportiness. Then there were those haunches which have always set Jaguars apart from the proletariat. And the curves over the headlights, like the bonnet had been poured into place, rather than bolted on.

Obviously, there are going to be drawbacks to any car that looks this good and the Jaguar was no exception. Inside, it was pretty cramped but somehow, that felt right. Some people like to feel they're driving around in an aircraft hanger, which is fine, but I prefer a hemmed in, cosseting feel and no car ever did that better than the Series Three. There was the high waistline and narrow windows so that it felt like you were driving a post box, peering at a silent outside world through a slot. And Jaguar was still using real wood and leather long after everyone else had switched over to petro-chemical lookalikes. Then there was the pencil tray. Right in the middle of the dash, there was a shelf, lined with velvet on which you kept pencils. And to ensure you could find them in a hurry, it was illuminated by a hidden bulb; as were the footwells. It was like a cottage in there with standard lamps and a wood burning stove for a heater. The doors, really, should have been lined with books. And yet, this 17th-century, thatched cottage could slice through the air at up to 150 mph, silently, just like the cat from which it takes its name.

Sometimes I see Series Three V12s for sale in the newspaper and I must confess, it's always tempting. They cost so little and yet they do so much.**jc**

Jaguar XJ6 Coupe

In the early 1970s Jaguar's XJ6 saloon had the look of Penelope Keith in The Good Life. Kind of sexy, aloof with little sense of humour. It was the saloon for serious-minded suburban bank managers. Yet Jag's reputation meant that there were scores of thrusting young executives with hairy chests and gold medallions who wanted a car with a cat on the bonnet. The E-Type was no more, the XJS was not quite to anybody's taste, and the saloon was just too dull to contemplate. And then the Avengers reappeared as the New Avengers, with Joanna Lumley driving a TR7 and Steed, having lost his chitty bang bang motor of the 1960s, speeding around Vancouver in a bright red, flared-arched, fat-tyred XJ6 ... coupe! At last, the saloon looked hip. To all intents and purposes Jaguar simply left off the rear doors, added a vinyl roof and at first used the 4.2 engine, but it worked (they later put the V12 in, as well). Most importantly, the coupe was pillarless, so when you rolled all the windows down, it looked positively American. Which was odd, because Jaguar never sold it over there.

It was only built for three years, and unfortunately those three years coincided with some of the worst car-building and management strategies in British motor manufacturing, so there aren't many left at all now, having rusted away with nationalisation and union power. And that's a shame, because it really was the nicest XJ6 Jaguar built.**jc**

> *To all intents and purposes Jaguar simply left off the rear doors*

When the Series Three Jaguar was replaced by the old XJ40, it was a sad day for those who appreciate the finer things in life. Yes, the new car was still very obviously a Jag, and a well-made one to boot, but the neat little details had been lost. It was like a cheetah without its spots. Still a cheetah but not quite right somehow. Well the spots are back with the new car, along with the haunches, the lumps of chrome and most importantly of all, the sculptured bonnet. This is all good news, but the best news of all is the supercharged XJR figurehead.

I've run one of these cars for a year now and can tell you this. No car on the planet today can do quite so much quite so well. Think of it as a quiet, comfortable means of taking four people and some suitcases to the south of France, and it will oblige.

Think of it as a means of getting doormen to stand to attention at swanky hotels, and it will oblige. Think of it as a means of getting respect in the dodgier parts of South London, and it will oblige. Think of it as a four-door Ferrari though, and it doesn't just oblige. It gets out there and tears your eyeballs out.

It handles, steers, stops and goes like a small sports car and I simply love it to death.**jc**

No car on the planet today can do quite so much quite so well

Jaguar XJ220

killer cat

Engine V6, 3500cc
BHP 542
Brakes .... To stop a Chieftan
Top speed .... 213 mph
0–60mph .... 3.6 seconds
MPG................. You're joking?
Looks .... Very, very fast
Today's cost £403,673

For about five glorious minutes, the Jaguar XJ220 with its top speed of 217 mph, was the fastest production car money could buy. But then along came the McLaren F1 which can top 230 mph and the Jag, instantly, was a relic.

Now you won't find an F1 in this book because I don't believe any car should cost more than a six-bedroomed house in the Cotswolds but the Jag is here because ... er, um. Well I don't know why exactly. I drove one only once in Dubai and can recognise that there are a huge number of shortcomings. It feels heavier than it really is, and it really is heavy. And

the cockpit is more cramped than a porn star's underpants. And you can't see out of the back properly. And worst of all, the Metro 6R4 engine, despite the fancy turbocharging, is still a Metro 6R4 engine – which means it's about 42 miles away from being a classic.

Nevertheless, overall, the 220 is a giggle mainly because even in Dubai where money grows in the desert, people stop and stare like you're from the planet Zarg. Not for long though because it really does move. The acceleration isn't electrifying as it is in, say, a Lotus Elise but the g forces just

keep on coming, and coming and coming. And then they come some more. And then, as you sail past 200 mph and there's no let up, you do too.

If I may liken it to an animal, I'd say it was a Brontosaurus. Yes, that's a good one because the XJ220 was a dinosaur and it's now extinct. And that's a teensy bit sad.**jc**

Jensen Interceptor

Jason King in Department S had the optional ice bucket and cheroot holder

In 1971, Jason King was TV's Mr Smooth. His shirts had ruffles, his droopy moustache almost met his sideburns, his hair covered his high, velvet collars. His shoes had cuban heels, his crushed velvet flares flapped around his skinny ankles and the huge rings on his long, elegant fingers flashed as he put his More menthol cigarette to his lipsticked lips. He lived in a state-of-the-art apartment with venetian blinds, remote control lights, curtains and stereo and "worked" as a spy.

His work seemed to involve dating foreign models with too much eyeliner and short skirts, shooting people with a little Luger, fighting with dark-suited foreigners and driving very, very fast along deserted London streets.

I seem to recall he did this in the only car of the time that suited him perfectly: the Jensen Interceptor. And if he didn't, he should have. Here was a car that dated as fast as his nylon, paisley-print matching tie and shirt, that was Space Age in that peculiarly 1970s manner, and although looked cumbersome, had the power of a rocket. Like King in fights, the Interceptor glided smoothly over bumps, but could kick like Bruce Lee.

The Interceptor had flared arches, cream leather seats (or all seemed to have, anyway) and a wrap-around windscreen. It was designed in Italy, built in Britain, and powered by a Chrysler V8, 6.3 litre engine. It was great for picking up women.**jc**

platform soul

Engine	V8, 6276cc
BHP	330
Brakes	discs all round
Top speed	135 mph
0–60mph	7 seconds
Weird bit	early ABS
When?	1966–1976
Today's cost	£11–£20,000

Lamborghini Miura

Over the years, Enzo Ferrari inspired a great many cars that wore the prancing horse badge with pride. And two that didn't wear the prancing horse badge at all. When he pulled out of a deal with Ford, Henry was so cross he went away, determined to build a car that would crush Ferrari on the race tracks. And he did just that with the GT40. Then there was Ferruccio Lamborghini, a wealthy tractor manufacturer, who hailed from the same region of Italy as

Enzo. He wanted to buy one of the scarlet screamers but this was a time, remember, when Ferrari was more of a racing team than a road car manufacturer so ordinary customers weren't afforded much respect.

Lamborghini was so appalled by his treatment that he decided to build a better car than Ferrari. His first efforts fell short of the mark by some way but at the Geneva motor show, in 1966, he took the wraps off a car that quite literally

stopped the world in its tracks – the Miura. Everyone had become used to mid-engined racing cars – the GT40 for example – but this one was a mid-engined car for the road. And not only that, but its massive V12 was mounted transversely. Then there was the body. Penned by Marcello Gandini of Bertone, it was by far and away the most futuristic shape ever to see the light of day; so futuristic in fact that even today, a Miura can still spin heads as easily as its big V12

We talk, glibly, about cars being able to fly. But this one really could

will spin those rear wheels.

Look at the pictures of it and remind yourself that we are talking about 1966 – a time when everyone had Ford Anglias. And now look at the pictures again and wonder if you've ever seen a better-looking car. Sadly, it was no means perfect to drive even by 1966 standards. Owners reported that as they got close to the 170 mph top speed, the front end would try to take off – but they were lying. I've driven a Miura and can report

that at just 80mph, the steering becomes eerily light as the front starts to point upwards. It seems there were no wind tunnels in those days and that by creating such a sleek shape, Gandini had inadvertently made something with the aerodynamic properties of a plane. We talk, glibly, about cars being able to fly. But this one really could.

And in addition, it was built with about as much care as you or I would put into a bird table. It was fantastically unreliable then, and it still is now. I had more breakdowns from the Miura in one day than I've ever had in 20 years of driving.

But these were small peccadilloes to the world's super rich who took the Miura to their hearts. It was very fast and it did handle with the sort of aplomb most thought was reserved only for the race track. Surprisingly, Enzo didn't react as quickly to the Miura as you might expect. His answer-back car was the Daytona – with its engine in the front.

But over the years, the Miura was treated to an endless stream of mechanical changes so that by 1973, the P400SV with 385 bhp on tap could crack 175 mph. And that made the Daytona look like a crippled slug. This really did wake Enzo up and he ordered his designers to give the world a mid-engined car to take on the Miura. They came up with the 512 Berlinetta Boxer, another beautiful and fast car but was it really a match for the Miura, I wonder? I've driven both and would say that to drive, the BB was superior, as well it

should be. It's ten years younger than the Miura. The BB doesn't take off, for instance and it has more grip too, though if you overcook it, you need the reactions of a fly to prevent a spin. With supercars though, driving is only half the appeal. Styling is just as important and on that front, the Miura is right up there as one of my favourite all-time designs.

The best version of them all is the Jota, which came along in 1969 as part of an aborted attempt to turn the road car into a serious racer. Somehow, they shaved 770 lbs off the weight and tweaked the engine so it delivered an almost unbelievable 440 bhp.

---

yes yes yes

Engine	**dohc, V12, 3929cc**
BHP	**350 ('S' 370, 'SV' 385)**
Brakes	**discs all round**
Top speed	**160–175+ mph**
0–60mph	**6.3–6.7 seconds**
When?	**1966–1973**
Looks	**The business**
Today's cost	**£60–£85,000**

Just days after Boeing made an abortive maiden flight with their new 747, Lamborghini took their latest Miura up to 190 mph. At that speed, I have no doubt the Miura would be a better plane than even the Jumbo. My day with a Miura was fraught from the moment I climbed inside. Every time it stopped, the plugs would foul and the engine would stall. And the battery didn't have enough juice to fire it up again.

So I'd have to find someone to give me a push which was embarrassing. Especially when you'd only get round the corner and at the next traffic lights, exactly the same thing would happen again. The steering was stupendously heavy, as was the clutch. And there was no air conditioning to combat the heat generated from either my exertions or that gigantic engine.

And the visibility was terrible too. Pull up at a slightly oblique junction and the thick, rear buttresses would stop you seeing if anything was coming. As you sat there for an instant, wondering what to do, the engine would cough once and die.

And when I did find a road more suited to the Miura, it played Piper Cherokee down every straight. It was, I think, the worst day I've ever spent with a car but with those looks and that wailing, chirping, bellowing, barking, snuffling engine, I'll always remember the car with a fondness it doesn't really deserve.**jc**

ghi

Lamborghini Espada

Looks like (but isn't) Ed Straker's car in UFO

Two years after producing the stunning Miura, Ferrucio Lamborghini presented the world with a new model which he just knew would upset Enzo Ferrari: the Espada.

A genuine four-seater GT with a four-litre V12 sitting up front under a long sleek bonnet, it pulled a stunning-looking, minimalist Bertone-styled body.

It took Ferrari four years to come up with a new, genuine 2+2 to rival the Espada, and even then the 400i couldn't match it for looks.

For ten years the Espada was undeniably the world's best looking 2+2 Italian GT, and it only lost the claim when it ceased production.

In 1968 it looked like the car that families would be driving in 2002. In fact, the prototype sported gullwing doors

and must have had a profound effect on whoever made the futuristic television series UFO. In 1978, when the last of the 1200 or so that were made rolled out of the factory it still looked fabulous, and hadn't suffered any significant style changes since the first production run. Which is not something that can be said about many cars.

And that includes Ferraris.**jc**

Lancia Fulvia HF

have fun

Engine....................	V4, 1584cc
BHP	130
Brakes..........	discs all round
Top speed.................	115 mph
0–60mph..............	8 seconds
When?	1968–1972
Today's cost	£11–£15,000

I've never been to the Galapagos Islands but I know it's hot there. I've never eaten a dog either but I know it will taste like chicken. Everything unusual tastes like chicken. I've eaten a snake, and it tastes like chicken. I've eaten kangaroo, cat, crocodile and rather burnt rice and slug, and they all taste like chicken. I've never been to a photocopier convention but I know it would be dull.

What I'm trying to say is that you don't need personal experience of something to have a pretty good idea what's in store. And that's why I feel confident about putting the Lancia Fulvia in this book. I've never driven one and to be

honest, apart from engine sizes and the fact it's front-wheel drive, I don't know very much about it. I don't know who designed it, or why Lancia decided to make it. I don't know if it was a rot box or a little gem.

But I do know that whenever I see one, I always say 'ooh look, a Fulvia'. I just like them, OK.**jc**

Lancia Stratos

I don't care how many times I'm told that Paris or New York is the epicentre of fashion, I know that ground zero is Milan. Experts say London is the current centre of cool but people still walk down Oxford Street with their bra straps showing. And that's not cool at all. It's revolting. In Milan, the catwalk queens are tabbies compared to the sheer slinkiness you find in any one of a thousand pavement cafés. It is a city where men only ever go grey round their temples and where women only ever cross their legs at the ankle. London? Cool? Do me a favour. Soon, people will be trying to argue that Germany or Detroit is the centre of automotive design. But again, that's hopelessly wrong because of this simple fact. Eighty per cent of all cars ever made were either designed in, or inspired by, a Torinese design house. The sheer number of great car designers based in Turin boggles the mind. And most started out in life working for Bertone which is best remembered for the diabolical Ferrari 308GT4.

That's unfair because Bertone's finest hour, without doubt was the Lancia Stratos. It started out in life as a concept car, designed to draw the crowds to his Turin motor show stand in 1970. Lancia's rally team boss liked what he saw and in that mad, panicky-type way that defines Italian motorsport, worked quickly to find an engine. He came up with the V6 from the Ferrari Dino. Bertone was asked to make 500 Stratoses (Strati?) which was the sort of awkward number that fell between hand-made craftsmanship and full-blown, mass production. Amazingly, the first car was ready in just a year and three years after that, it won the world rally championship. Something it did again in 1975. And again in 1976.

I drove a 140 mph, road-going version and found it to be one of the trickiest, most brutal sons of bitches ever to take to the roads. You sit, almost in the middle with the pedals off to the left and the wheel to your right. Imagine trying to drive while facing sideways and you'll be near the mark but now think about ferocious acceleration and an incredibly short wheel base which causes monumental oversteer to arrive without warning. Mix the resultant fear with heat from engine, remember there are no wind-down windows, and you have the

world's first supersonic sauna. I loved it. Ferrari's contribution was great. Lancia's decision to make the car was to be admired, and the drivers who took it to victory have imparted a sense of history. But the best thing about a Stratos is the way it looks.

Yes I know taste somehow by-passed the 1970s and that, as a result, the Stratos came in quite the wildest paint schemes seen anywhere outside a graffiti convention under the A40 flyover in Shepherd's Bush. But never forget that even a lime-green Stratos is still a Stratos.

And as a result it is still a cocktail of everything at which Italy shines.**JC**

Never forget that even a lime-green Stratos is still a Stratos

Lancia Stratos

Looks like every radio-control toy car you can buy

Lancia Delta Integrale

So what's the most boring sport on earth? I see the battle as being a two-corned contest between cricket (the only game so dull, that players have to stop for tea) and rallying. I suspect rallying is more fun to do but as a spectator, it is absolutely hopeless. It is cold, dark and muddy and every time a car goes by, it spews a bucketful of small stones into your face.

At least if you're forced to watch cricket you can nod off but this is not possible on a sheet of plastic at 2 am in the middle of a sub-zero forest in Wales. However, what swings it is that rallying does have one redeeming feature. The cars that compete must be available to buy, and that means Lancia was

forced to sell us the incredible Integrale.

And as the rally car evolved, so did the road car which meant that every year, the grip and power just got better and better. Overall, it's probably superior to its chief rival, the Escort Cosworth but then again, the Ford had its steering wheel on the right in Britain and the Lancia didn't. And there's no point having all that power and all that grip if you can't see past the Marina that's holding you up.

However, find an open road, put something loud on the stereo to mask the squeaks and rattles and you probably won't find a more complete all rounder.**jc**

Lancia Monte Carlo

This, like the Delta, was also turned into an all-conquering rally weapon, but it is the ordinary 2.0 litre car that I've chosen.

First of all, it looks like it should have cost Ferrari money but it was less than £9,000 new. And for that you got a 120 bhp mid-mounted engine, and a snug cockpit which featured just about the best seat upholstery of all time.

This was an affordable supercar but it was let down badly by its brakes. Early models, which can be identified by solid rear buttresses would lock their front wheels if you so much as blew on the pedal. The problem was so bad Lancia took the Monte Carlo out of production for a couple of years ... and then brought it back again. The only time that I can ever remember this happening.

And so what had they done to cure the problem? Simple. They'd disconnected the servo. Amazingly, it worked. Yes, you had to give the pedal more of a shove but at least you could brake and steer at the same time.**jc**

Just about the best seat upholstery of all time

Lotus Elan

In 1962, Colin Chapman's Lotus unveiled a new sportscar that he hoped would be the making of his company. His first attempt at making an affordable, fast, light, British sportscar had not been entirely successful, the Elite being just too expensive to make work. For the Elan he dropped the Coventry Climax engine, and used instead a rugged, commonplace Ford four-cylinder 1600 block. Being Lotus though, Chapman couldn't allow a dull old Ford engine to drive his car, so he had his people develop their own twincam head for it, and presto! A new, cheap, fast and extremely good-looking British sportscar that would become the ultimate object of desire for successive generations. Mazda liked it so much they based their MX5 on it.

The Elan, with its hidden lights, round taillights and simple sporting dashboard, is simply chic. And, since the moment that Honor Blackman, dressed completely in black leather, leaped out of hers in an episode of The Avengers, the Elan has been simply sexy.

Later models would have half a rear seat, longer wheelbase and special rating taking the BHP from 105 to 126, but the convertible first series (the so-called Baby Elan) is the ultimate Elan. It might look like Emma Peel's car, but it handles corners better than any number of modern macho motors, and is a lot more fun, too.**jc**

so a-Peel-ing

Engine.....	dohc, 4 cyl, 1558cc
BHP	105
Brakes..........	discs all round
Top speed..................	115 mph
0–60mph	8.5 seconds
When?	1962–1967
How many?	9659
Looks	Emma Peel
Today's cost	£10–£15,000

Lotus Esprit Turbo

By the end of the 1970s, Colin Chapman's warm embracing of the Common Market, which led to a 1400cc Renault engine being put into the first Europa models, had grown to the point where he gave the task of designing his new, non-kit form, mid-engined coupe, the Esprit, to Italian styling guru Giugiaro and his Ital team. (Yup, the same ones who turned the Marina into that lovely saloon the Ital for BL.) The front-engined Elite and Eclat, for all their marvellous 1970s kitsch appeal, had failed to conquer the motoring world so the new Esprit was being depended on. Unfortunately the first examples suffered badly from the shakes at any speed, and

that speed only got close to 135mph, which was not, by any reasonable standard, rip-roaring stuff. It also looked like a piece of funny-coloured cheese with wheels.

Something, then, had to be done. And it was. For his next film, The Spy Who Loved Me, James Bond needed a new British supercar to replace the ageing DB5, and Lotus had the task of providing him with it. Perhaps it was M who suggested putting a turbo charger on to the 2.2-litre engine, thus upgrading the BHP from a middling 160, to a more respectable 210 and powering the V-shaped wedge on wheels to 150mph, and from 0-60 in just over six seconds.

the name's Bond

Engine	dohc, 4 cyl, 2174cc
BHP	210
Brakes	discs all round
Top speed	150 mph
0–60mph	6.1 seconds
When?	1980–1981
How many?	Not a lot
Today's cost	£10,000

Whoever it was, that man should be congratulated. At last Lotus had a very fast, not bad-looking car with supercar performance, at a significantly less than supercar price.

So Lotus was saved. At least for a few months, anyway.**jc**

Lotus Elise

When I first heard that Lotus was planning to build an all-new sports car, I laughed so much, that I burst.

Here was a company whose founder and mentor was long since dead. It had been rescued from the receivers by General Motors who hadn't a clue what to do with their new toy and who passed it on, like a hot hand grenade, to Bugatti – a mysterious new Italian company that went bust.

Financially therefore Lotus was about as stable as the San Andreas fault.

And what about the cars? Well, in the early days, Lotus stood for Lots Of Trouble, Usually Serious but then, along had come the Esprit. James Bond used one in the Spy Who Loved Me and again in For Your Eyes Only. Then there was another Esprit which was about the same as the first.

They had turbocharged and developed it as fully as seemed possible. Then there was a new turbo Esprit, followed quickly by another, and then, to everyone's surprise, another. A Turbo Esprit saw the company into 1989, and 1990, and 1991. At some stage, there was a new Elan but that didn't work so they went back to making Esprits again.

And then they made the new Elan again but it didn't work second time round either, so they were left with the Esprit, no Grand Prix team and no money.

How could a company with this kind of track record possibly hope to make a new car? And where on earth would they find the cash to get it into production? Well they did make the car and they were bought by Proton and now the Elise is with us, serving as a constant reminder that he who laughs last laughs longest.

This is perhaps the most extraordinary little sports car I've ever encountered, and now I'm going to drop the word

'perhaps' and with it any shred of doubt. This is THE best driver's car in the world.

Some will tell you about its extruded construction methods. Some may point to the technically interesting Rover K series engine that propels it. Others may note the interesting shape, or the wonkiness of the hood or the lack of carpets.

But one thing is a thousand times more important than any of this stuff – its handling. With its ultra-light weight, its engine in the middle and the wheels pushed to the very corners of the car, it is as controllable as the best of breed at Crufts. I can think of a thousand cars with more grip, and consequently more road holding but none which rewards and responds quite like an Elise. And anyway, when a Ferrari lets go at 110 mph, it's scary. Round the same corner, an Elise would let go at probably 60, which means your corrective decisions aren't made under a sheen of fear.

You ease off a bit, steer into the slide and feel the precise moment when the car begins to grip. It is text book stuff so you won't over correct and face miles of trying to tame the fishes tail. Only a complete clot would ever lose an Elise, in the same way that only a complete fool would ever buy anything else.

That said, it is hopeless for long distance work, and its roof is stupidly difficult, but for a quick blast, it cannot be bettered.

And so far, I've only seen one at the side of the road in a cloud of steam.**jc**

Lotus Elise

Maserati Sebring

The worst car I've ever tested as a motoring journalist was the Maserati Quattroporte. When I was a small and eager boy, I remember my father saying he wanted a nanny who'd be prepared to clear out the gutters and lend him her Maserati at the weekend. Not a Ferrari, you'll note, or an Aston Martin. He wanted a Maserati because even as late as the mid-1960s, Mazzers were something else. They were the IT cars of their time. Maserati had been racing since the days of Ben Hur and had built up a formidable reputation.

But, after the war, they began to think a little bit more seriously about their road cars. In racing, Maserati made any number of engines but for the most part, their road cars used a rugged in-line six. Yes, the Sebring could do 150 mph but obviously it was no match for the V12 Ferraris. A valid point but Maserati isn't Ferrari. At this time, their road cars were different. Visualise a Ferrari and it's charging up a mountain pass, with its blood-curdling V12 exhaust bouncing off the granite walls and its squealing tyres providing the descant. But a Maserati, though equally sporty and revered, is more at home tooling along the Croisette in Cannes, on its way to pick up Kirk

Douglas and take him for dinner on Gunther Sach's boat in Monaco harbour.

So why the Sebring then? Well this, I reckon, was the last and therefore the most advanced of the proper Maseratis. Afterwards, they moved into an entirely different territory – the land of the supercar. The Sebring was an honest-to-goodness GT car. And it had what no Maserati has today – style. By the bucket load.**jc**

big smoothy

Engine ...	dohc, straight-six,
Brakes	discs all round
Top speed	135 mph
0–60mph	8.4 seconds
When?	1962–1965
How many?	346
Today's cost	£20,000

Maserati Ghibli

At the age of 13, I was taken away to boarding school in my father's Audi. And this was a big concern.

I was a kid from Doncaster being parcelled off to a public school where I figured, everyone would be a multi-millionaire. They would laugh when they saw my father's Audi swing through the school gates.

And my worst fears were realised when he did just that and there, right in front of the imposing door, was a Maserati Ghibli. I'd never seen anything quite like it. It was the pointiest, sleekest-looking car I'd ever actually seen outside a motor show. And it was huge; so huge in fact that there was room inside for a school trunk.

Here was a car that could get from 0 to 60 in 7.5 seconds and which could reach 155 mph. It brings shivers to my spine, even now. And it would bring shivers to yours too, if you ever drove one because under the amazing body, it had cart springs which meant it handled and rode like a Standard Vanguard. **jc**

Under the amazing body, it had cart springs

85

Maserati Khamsin

In 1973, Maserati replaced the Ghibli with the Khamsin. They were firmly entrenched by then as a supercar maker and realised that sleek looks alone were not enough. The Khamsin needed to handle as well as it turned heads. So, though it had the same 4.9-litre V8 as the Ghibli, it was a completely different car underneath. Maserati, at this time, was owned by Citroën who foisted their unusual hydropneumatic systems on the Italians ... and it worked. The Khamsin was handsome, fast, fun to drive and it had a racing pedigree to beat even Ferrari. This was one of the all-time great supercars. I just wish Maserati would make something like it today. **jc**

Italo-French relations

Engine	dohc, V8, 4930cc
BHP	320
Brakes	by Citroën
Steering	ditto
Top speed	160 mph
0–60mph	8.1 seconds
When?	1974–1982
How many?	421
Looks	by Bertone
Today's cost	£17,000

Matra Murena

For some reason, during the 1970s, people got it into their head that it would be a very good idea to have cars in which the steering wheel was neither on the left or right hand side, but in the middle. Science fiction writers seemed to love the idea. Car manufacturers would probably love it too, since it meant that you'd have a standard design template, and could build cars for both the British and American market at the same time, on the same production line.

Which is all well and good, until you think about the matter of overtaking. Without a passenger either side of the driver to assist, you would have to get into the middle of the road before deciding whether to get into the middle of the road or not, to overtake.

However, French car makers Matra decided that the three seats with driver in the middle was a great idea, and so built a neat little wedge-shaped sports car with just that. It was fun. It was deemed illegal over here. So they moved the wheel across to the left, and kept the three seats in a row. Years later McLaren built the F1 with a steering wheel in the middle, but that's fast enough to overtake with your eyes closed. The Murena wasn't.**jc**

Mazda RX7

Blah blah Wankel rotary engine blah blah blah so smooth that a buzzer sounds when it's time to change up blah blah blah 21st century technology. Yes, a Wankel rotary engine is smooth but it is also thirsty and if it's so great, how come everyone isn't going for it, huh? I think it's a gimmick actually and it would be the very last reason on earth why I'd choose to buy a Mazda RX7. I would however buy one because it looks utterly wonderful; like an E- Type Jag for the 1990s. And because it has traditional white on black instruments which are surrounded by chrome.

That's it. That's enough.**jc**

Mercedes 600 Pullman

It is widely accepted that in the 1960s, bad people drove Mark Two Jaguars. But these guys were the foot soldiers, operating at the front line with their sawn-off shooters and broken noses. These guys were dispensible.

Somewhere in London's East End, they were being controlled by the Bob Hoskins character; a seemingly legitimate businessman who organised the armed robberies and provided the shooters and the Jags. This guy would have had a Mercedes 600 Pullman – the largest and most sumptuous car that money could buy in those heady days of Mary Quant and Carnaby Street. It had a 6.9-litre V8 engine but this was so silent, you could almost convince yourself that it was actually being powered by dylithium crystals. It's an engine Jim, but not as we know it. Certainly as Mr Big

luxuriated in the back, he was untroubled by either sound or the sensation of being in contact with planet earth – the 600 appeared to hover just above the road to ensure that people inside never wobbled or vibrated. Might have been dangerous. Their guns could have gone off. Or worse, they could have spilled their drugs. In keeping with the bad boy image of rock'n' roll, the music world also embraced the 600 to the point where every Beatle bar Paul McCartney had one. As did David Bowie.**jc**

It's an engine Jim, but not as we know it

Mercedes 500SL

Ever since the dawn of automotive design, we've been able to take it for granted that Mercs are well made and solid, and that they don't depreciate much. And that they're safe, and fairly swift and robust. But the SL adds a touch of fire water to the mix because it is about as stylish as cars get. The trouble I've faced here has been to choose which one of these German wunderwagons to list.

Well the first is too old, so that one was an early casualty. The third had its image all messed up by Bobby Ewing who used one in Dallas, surging out of the driveway every week so that he could go off and be nice to his hairdresser. It's hard to take a man seriously when he spent four years in the shower. And if you don't like

the man, then you have to suspect his wheels too. So that leaves the second version which was pretty as a picture, and the fourth, the current one, and the model I've selected. I've said, elsewhere in this book, that no car on the planet does quite so much quite so well as a Jaguar XJR, but the Mercedes 500SL gets awfully close. Yes, it can only take two people whereas the Jag can take four, but in the SL, you push a button and 11 electric motors get to work, packing the roof away in a cubby hole behind the cockpit.

Then, you have the wind in your hair but even at 155 mph, it never becomes a hurricane because of the deflector which sits behind the occupants' heads. Nor does it become dangerous because if the

a place in the sun

Engine V8, 4973cc
BHP..................................... 308
Brakes.......... discs all round
Top speed .... 153 mph
0–60mph .... 7.5 seconds
Today's cost............. £81, 390

car starts to roll over, a roll bar flicks up to save your hairstyle from any advancing leaves. It's fast and exciting and a delightful car to behold and yet, you can never forget that it's a Mercedes because it's safe, solid and robust too. And it depreciates slowly and so on, and so on, and so on. Basically, it's one of the very few cars out there that you can buy with your heart, and your head.**jc**

Mercedes S600

So, if I'm such a fan of the Jaguar XJR, what pray is this German interloper doing here? And why do I keep insisting on referring to it as the best car in the world? Well that's simple. It is.

There are prettier cars, and faster cars and cars that handle better but none do what their designers intended quite so well as the Mercedes 600S. And what its designers intended, by the way, was to make the best car in the world.

It is a whopper but none of the exterior bulk is wasted which becomes apparent when you climb into the back and shut the door. DON'T slam it – because as soon as it is nearly closed, a small electric motor swings into operation and gently pulls it shut.

You settle into your seat and, even if you're 6'5" like me, you can stretch out, and settle back wondering if that faint hum coming from miles away in the front means the engine has started.

It has, and that has brought your various toys to life: the drop-down, illuminated mirrors in the roof, the remote-control device to operate the front-mounted stereo, the electric-seat, backrest adjustment, the electric headrests, the electric rear roller blinds and your own air conditioning controls. And remember, this is just the back.

In the front, it gets even more amazing not just because of the toys but because of the way this tank picks up its iron petticoat and flies. The massive 6.0-litre V12 engine, though not quite as silent as Jaguar's 5.3, is nevertheless a fair old

fist. Put your foot down and fatty in the back will spill his drink, that's for sure.

To try and keep the two-ton monster from ploughing straight on at every bend, it has as many sensors underneath as the Nissan Skyline; only they're there for safety reasons rather than a means of going more quickly. Turn into a corner too fast and you can feel the car braking whichever wheel needs braking. And redistributing power as it tries to work out if the driver's gone mad or not.

I once hurled one of these limos round Mallory Park race circuit and it got so cross, I half expected a boxing glove to spring out of the sun visor to teach me a lesson. So no, it's nowhere near as much fun to drive as an XJR.

But this has radar sensors front and back to make parking easier. Early models had little aerials that slid out of the rear corners every time you engaged reverse, so that you could tell where the back of the car was. And this was the first car I ever drove that automatically shut its double-glazed windows and sunroof when you locked the doors.

Then there's the wood on the dashboard, taken from the already dead root of trees that are specially grown by Mercedes anyway. The offcuts from the windows are used to make milk bottles and the factory where the S class is made uses no water. It just recycles the same stuff over and over again.

It is attention to detail like that which makes the 600 stand out from the crowd. Yes a Lexus is quieter and yes a BMW has more gimmicks and yes, a Rolls Royce has thicker carpets, but generally speaking, corporate high flyers – the people who can afford to spend £100,000 on a car – only want the last word in reliability, refinement and safety.

They want to feel, when they're in their car, that they're actually in a suite at the Dorchester. There's only one answer really. **jc**

cruiseliner

Engine	V12, 5987cc
BHP	389
Top speed	155 mph
0–60mph	6.6 seconds
Today's cost	£102, 490

Mini Cooper

It would, of course, be possible to write an entire encyclopaedia on the Mini, and over the years a great many people with beards and dirty finger nails have done just that. The ordinary cooking version was a clever little car for its time but that time was 40 years ago, and in the same way that they don't play Status Quo on Radio One any more, I'm not going to sing its praises here. But the Cooper was different. Oh sure, it's not fast and it is very definitely not comfortable. It handles well but you can do better and you can certainly get a lot more practicality as well.

However, thanks, I think to the Italian Job, the Cooper is a legend which means age only serves to enhance its reputation. Look at it this way – when you think of Jimi Hendrix, you think of a guitar genius, a man ahead of his time, the figurehead of a generation. But never forget that he choked on his own vomit.**jc**

Min

It's not fast and it is very definitely not comfortable

Costs more Swiss francs than you can carry in its spacious boot

Engine	V8, 7210cc /6974cc
BHP	375/450
Brakes..........	discs all round
Top speed	152 mph
0–60mph	6.3 seconds
Rare as?	Hens' teeth

Switzerland: big on cuckoo clocks and banks but not so big when it comes to cars. Indeed, Switzerland is the most anti-car country on earth and will, by the turn of the century, be nothing more than a hilly bus lane. Such is the Swiss distaste for all things automotive that motor racing is actually banned. But there was once a half-hearted attempt to give the greenest country on earth a car industry. In 1967, a chap called Peter Monteverdi announced that he was going to take on Germany and Italy and Britain with a new grand tourer. And the Monteverdi motor car was born. But the actual Swiss involvement was minimal. The bodies were designed and built in Italy which is why it was so handsome. The Chrysler V8 came from America, which is why it was so powerful and the suspension was copied from Aston Martin which is why it handled so well. It was a superb car; so superb that Monteverdi applied a massive price tag which kept production down to about one car a week. That's not really enough and 10 years later he packed it in, and handed Switzerland back to the eco-weenies.**jc**

Nissan Skyline

Nissan Skyline

When a car starts to oversteer, I know exactly what you're supposed to do. But I'm usually so scared that I simply undo my seatbelt and climb in the back. There are cars, like the Lotus Elise, which begin to slide at a fairly slow speed and that's great, but most high performance machinery grips until it's doing mach three. So when it lets go, you're sideways in a nanosecond and in a hedge shortly thereafter. The Nissan Skyline, however, is different. It has big, fat, grippy tyres and four-wheel drive so that it clings on as though its very life depends on it. You can take it round corners at a speed that simply defies belief. I mean so fast that your liver bursts through your rib cage and splatters into the door lining. So you'd think that when the laws of physics wake up and cause a spin, the poor old liver-less driver would be powerless to react fast enough. But this isn't so. When the rear tyres let go, the fastest acting computer outside the space shuttle sends power instantly to those at the front. There's nothing unusual in a torque sensing differential which can do the same trick but the Nissan does it electronically. And computers react far faster than fluid or drivers. So you turn into a corner far too fast and accelerate far too soon. In any ordinary car, this would be a recipe for disaster but a little dial on the dash, which indicates how much grunt is being sent to what part of the car, shows that you're caught in a war between physics and computers. And the computer always wins.

I haven't even got my seatbelt undone before the car has sorted itself out and is waiting for instructions about how fast it should tackle the next straight. And therein lies the Skyline's next party piece. The standard 2.6-litre turbo motor delivers less than 300 bhp, because Japanese law doesn't permit any more than that. But a simple tuning operation takes the car up to a massive 370 bhp – about the same as a Ferrari 355. So you have the four-wheel drive, the four-wheel steering and those computers to help you through the bends, and Himalayan power to get you down the straights. Small wonder that a Nissan Skyline is the only production car in the world to get round Germany's fearsome Nurburgring in less than eight minutes. I honestly believe that as a driver's car, it can be mentioned in the same breath as the Lotus Elise and the 355. And I'm fond too, of the way Nissan's engineers went about it. For years, Japan has simply copied European inventions, rather badly, but with the Skyline, they went their own way, using electronics to overcome dynamic shortfalls. And because Japan is so good at electronics, it works with devastating effect. I love the styling too. Again, it must have been tempting to wrap the wires in a sleek body, as Honda did with the NSX, but instead, Nissan simply fitted a body that would pass for a Tokyo taxi cab.

The new version is bedecked with spoilers and leery wheels but the original, and to my mind the best, could pass as a cardboard box. There were simply no outward clues that this was a serious car. It was the same story on the inside. You sat on brushed-nylon, pleblon seats staring at a vulgalour dash. Even by Nissan's own low, low standards, this was poor. But at least you had a thick, leather wheel and the aforementioned torque dial to remind you that nothing on the road will corner faster.

I'm such a fan of this car, the British Owner's Club have made me honorary president which is fine and flattering but I don't actually have one. I do however ensure that whenever I go to that soulless sprawl they call Tokyo, I borrow one from Nissan. It makes an unbearable country almost bearable. In a great country like Britain, it makes life pretty much perfect.**jc**

When a car starts to oversteer
I'm usually so scared that I
simply undo my seatbelt and
climb in the back

Nissan 300ZX

After Nissan killed off its 240Z, they managed to get the whole idea of sports coupes entirely wrong for the next 12 years. The 260 was worse than its predecessor, while the 280 was worse than anything that had ever gone before. Including Ghengis Khan. But even it was not as bad as the first incarnation of the 300. Anyone who knows anything about cars should always put this in their top ten worst cars of all time.

It was built for blue-rinsed, fat Americans so that when they ambled down to the shops, they could convince themselves that they were young and sporty. A quick glance at their trouserwear, however, would have reminded them that they were not. A quick dab on the car's throttle would have reminded them that the car wasn't either. So when I was asked to drive the new, 1990s 300ZX, my hopes were somewhere between the carpet and the floor. And as with all things, when you're not expecting much, anything is gratefully received.

In fact, the Nissan wouldn't have been a disappointment even if it had been based on a Tom Clancy book and starred Harrison Ford. With its twin-turbo engine, there was prodigious power but it was the smoothness that impressed most of all. It felt, in many ways, like it was a turbine. And the handling showed big car purpose too. At the Snetterton race track, there was wallow and roll but at no time did it ever do anything untoward. I always knew what was coming next, so I could plan, and be ready to unleash the horsepower ready for the next straight. And here's the best bit of all. I could even fit inside while wearing a helmet and that's a rare thing in any car, let alone a low-slung, grand tourer. Had this been impossible, it wouldn't have been the end of the world though because the roof panels lift out to make it a sort of convertible. I know it was designed almost exclusively for the American market but so, remember, was the Big Mac, and I could still eat eight of those in one sitting. No, it may have been a little out of its depth on British roads and it was a far cry from the original and wonderful 240Z, but I must confess that when EU regulations killed it off I shed a tear. However, I do hope that when Nissan get round to making another Z car, they have a look, very carefully, at what made the original 240 such a hit, and not at what average, lard-arsed Americans say in market research questionnaires.**jc**

Oldsmobile Cutlass

I admit that I know next to nothing about this car, except that as a small boy, watching The Man From U.N.C.L.E. I used to love how the plainly unreal background would rush past the windows of the big, black car that Napoleon Solo and Ilya Kuryakin drove around in. They'd have shoot-outs in that car, indulge in witty banter while driving away from a successful mission, outrun all manner of foreign agents fancy foreign saloons, and all the while not have to turn that massive steering wheel more than a third of an inch to make the tyres squeal.

When I was given a toy Man From U.N.C.L.E. car, complete with Solo leaning out of the back window firing his gun as you pushed it around the carpet, I eagerly turned the model over and read on the bottom the words Oldsmobile Cutlass Supreme.

At that point I'd never heard of Oldsmobile, and later in life when I really knew about cars, that fact didn't seem to need changing since most of their models seem very dull. But as a small boy, for the space of oh, several days, Oldsmobile was my favourite car manufacturer.**jc**

Opel Manta

You might by now have noticed that there are no Vauxhalls in this book. Which isn't really suprising, when you think about it. I toyed with the idea of putting in the HSR Chevette, which really did go like stink, but then, it looked so awful it was impossible to include.

Nothing built by the British arm of GM could ever shove its way into a book of Hot 100 cars. The German arm is different, though. In the early 1970s Opel were making a couple of coupes which both looked good, handled well and deserved the advertising slogan they carried, 'You can trust an Opel'. The GT

was a very pretty little sports coupe which was unfortunately let down by the Ascona 1.9-litre engine which didn't lift its performance to anywhere near the levels the looks demanded (0-60 in over 10 seconds, top speed of 115mph). Mind you, that's a whole lot better than the original 1.1 Kadett engine with which it was originally fitted.

The Manta GT was a much happier affair, though. Using the same engine in a good-sized, four-seater body, displaying Ferrari-style round rear lights and a curvy wedge shape, it was the Capri of choice for men with Italian aspirations

and Vauxhall money. Significantly, not many of the 100,000 GTs made still exist, and their current value is not much higher than that of a good Manta. After five years in production, for some strange reason Opel remodelled the Manta on a Capri, losing the pretty profile and taillights, making it look like a kind of Cavalier coupe.

It was probably on the demand of Vauxhall, who carried on making square, easy-rot grot-boxes, their equivalent sporting coupe to the early Manta being the droop-snoop Firenza which looked and drove like a sporting tractor.**jc**

Look at the list of cars I've actually owned since I started doing this motoring journalism thing. A VW Scirocco GTi. A BMW 3.0 CSL. A Honda CRX. An Alfa Romeo GTV6 and a Ferrari 355. You'll note a common thread. Practicality is not a big consideration. When I buy a car it need only be very good looking and blessed with a wonderful engine. That's it. Nothing else matters. But there is one car missing from that list; a car that I bought – by mistake actually – while covering the Paris Dakar rally in 1989. It was not good looking and its engine was not especially musical when it was new. When I came along 10 years later, it was about as

Peugeot 504 Estate

A car that I bought - by mistake actually

tuneful as Bjork. When you arrive at an out-of-the-way desert outpost in, say Niger, you will be surrounded by hordes of chaps wanting to sell you anything, and I do mean anything.

What I needed though was a car and a driver for a couple of days; someone who'd be at my beck and call and who knew the area well, someone who'd take me from the overnight rest halt to a good photographic location the next morning. It was the man I wanted most of all, rather than the machine and it was on that basis that I opened negotiations. Now I don't speak French terribly well. Even in France, they have trouble understanding my simplest stories about how I borrowed the pen of my aunt. But in West Africa, they speak a sort of French that is way beyond even a Professor of French at the Sorbonne. Nevertheless, I ploughed on, demanding that they clean up their areas with a lemming and a pick axe. After ten minutes, a chap grabbed my hand, took some money from my outstretched paw and gave me, in exchange, the keys to his Peugeot 504 estate car. It seems that for £7, I'd bought rather more than a taxi for two days. I'd bought the car. And I guess I still own it today. There's no doubt it'll still be out there because that is the way with the 504 estate. They cannot be killed. You could wind one up to 100 mph, though you'd need a tail wind, and drive it straight into a mountain, and the mountain would fall over. In crash tests, when they rammed the car into a concrete block to test the survivability of those inside, the concrete block simply shattered and the car emerged unscathed, as did the seven people inside. There's been a lot of talk about people carriers of late, though not in this book because I loathe them, but the 504 was doing all that school run stuff years before. And it was doing it in Africa too, in terrain that would kill a Land Rover in eight seconds flat. Go to the Sahara and you'll find it is still the most popular car even though it hasn't been made for years. They were shipped out there, and even though not one of them was ever serviced, they're all still going strong.

They run them on a mixture of paraffin and Mazola. They're driven where there are no roads, by people who don't know how to drive. And they're still out there, and they will be for ever. Even if there were a nuclear holocaust and everything on earth was wiped out, visiting aliens in a thousand years from now find that every 504 Estate had survived. And that they'd still start. For being the toughest car ever made, the Pug can wear its slot in the Hot 100 with pride. jc

Peugeot 504 Convertible

While you could expect the 504 estate to survive a nuclear holocaust, the convertible coupe of the range couldn't survive a car wash.

It looks big and strong, and for the first couple of years of its life, each car probably is, but once the air comes into contact with the bodyshell of a 504 convertible, the paint starts to blister and the metal turns to brown. So why is it here? I'll tell you. The 504 cabriolet is a very good-looking car indeed. And the V6 version was quite quick, too.

It's also a big car, and can carry four adults with ease, allowing all of them to lean seductively over the bodywork as it glides along the Croisette. Which is where its Pininfarina-styled curves belong. This is possibly the best-looking French cabriolet made in the 1970s. Which, of course, is where its problems lay. Why, during the decade that gave us Concorde, could mass car manufacturers not build machines to last longer than the average David Cassidy hit? Luckily, since Peugeot continued to build them until 1983, there are still a few examples of the 504 ragtop around, probably because they endear owners to spend all their time and savings keeping them warm, secure and rust-free.

Which mostly works, but because the car is so irresistible, people will insist on driving it to show off, even in the rain.

Unfortunately, they also insist on cleaning it ...**jc**

Peugeot 205GTi

Of all the people most likely to steal Mike Tyson's crown as the greatest heavyweight boxer of all time, Anthea Turner would seem to be an outside bet.

But stranger things have happened. Stranger things did happen in 1986 when the Volkswagen Golf GTi lost its title as undisputed hot hatchback champion of the world.

You might have expected the challenger to have come from Ford or Fiat but Peugeot? Peugeot had about as much to do with performance motoring as watercress. Peugeot made 504s to take on and beat Africa. They did not make pocket rockets to meet, leave alone beat, the all-conquering Golf.

Their first attempt failed. The 1.6-litre 205GTi was good but the 1.9 was something else altogether. It was faster than the Golf, better looking than the Golf, smaller than the Golf and at least as much fun to drive which meant that in every area that mattered it was at least an equal for the car that started it all.

It was so good in fact that for large chunks of the late 1980s, over half the cars in Fulham and Chelsea's streets were powerful Pugs. People used to refer to them simply as 1.9s and everyone knew what was meant.

It's been beaten now by its own little sister but back then, it was king of the hill, as important as a Puffa jacket and as fast as a Knopfler guitar solo.**jc**

Peugeot 106GTi

barking Pug

Engine	4 cylinder, 1587cc
BHP	120
Top speed	130 mph
0–60mph	8 seconds
Today's cost	£12,835

I've always been a fan out of the hot hatchback because it can do two things at the same time. It is enormous fun to drive AND it is practical enough to handle four people and some shopping.

I don't think I've ever driven a bad hot hatch, though the Maestro turbo comes close, which makes the job of picking out the best one an absolute nightmare. However, after much soul searching, I've decided to go for the Peugeot 106GTi; an F-15 fighter in a world full of Zeros.

Obviously, it's basically a standard Peugeot 106 so you get the benefit of space inside for a family, providing of course you don't have two sons who are line-out stars in the England rugby team, and a wife who eats too much.

There is a boot too which is accessed under a conventional hatch, and the rear seats fold down so you can use it for picking up awkward bits of self-assembly furniture. Sure, it only has three doors, but that aside, it is about as practical as a small car can be. And it is small too which means it's right at home squeezing through city traffic. Plus, if you get it wrong and take both front wings out while trying for a gap that just isn't there, they're cheap and easy to replace at any one of Peugeot's 2 million dealerships.

However, despite the healthy dollop of common sense, it's a fine-looking little car. There's a temptation these days to give urban runarounds a cheeky, curvy look but I think it simply makes them look idiotic. There's no doubt that the Nissan Micra is a good car but would you drive something that looks like it was squeezed out of a tube? No of course you wouldn't.

So let's say the designer decides to steer clear of the Noddy look, and remains faithful to a blend of curves and straight edges. If he's not careful he could end up with the Toyota Corolla and that's even worse. The people at Peugeot, however, managed to make the 106 look good by making it look bigger than it really is. And then, when they turned it into the GTI with those flared wheel arches, it adopted a menacing stance. Stroke a Micra and it will purr. Stroke this and it'll bite your bloody hand off.

And we're not talking here about a dog that barks its head off while reversing timidly into its kennel. Oh no. This little rocketship comes out of the traps like a greyhound which, only moments before, had had an entire jar of mustard poured onto its genitals.

Despite the modern day burdens of clean exhaust emissions, the Pug bounds past 60 in 8 seconds and keeps on bounding all the way to 130 mph.

This, of course is good, but what sets the 106 apart is the way it will attack any corner at seemingly any speed. You just turn the wheel, lift a little to bring the back end round and then nail the throttle again and it screams round.

Ordinarily, this would be unnerving but in a 106 it's like having a bath ... with Claudia Schiffer. Unnerving? No. Exciting? Oh yes.

Someone at Peugeot understands suspension. He showed this first with the 205 and he's been at it ever since, blessing every Pug for the last 15 years with a blend of ride comfort and handling that no other ordinary car can match.

And he can tune the car to suit the mood perfectly. If it's a 406 saloon, more emphasis is placed on comfort but the handling is still good. While in the 106GTI, there's all the road holding you could ever need, but the comfort is still there too. It'll never be remembered as a great car because it's too common and too disposable. Hot hatches are fashion accessories where it's important to have the latest and the best. Well right now, this is where it's at and you simply can't do better.**jc**

However, despite the healthy dollop of common sense, it's a fine-looking little car

Peugeot 106GTi

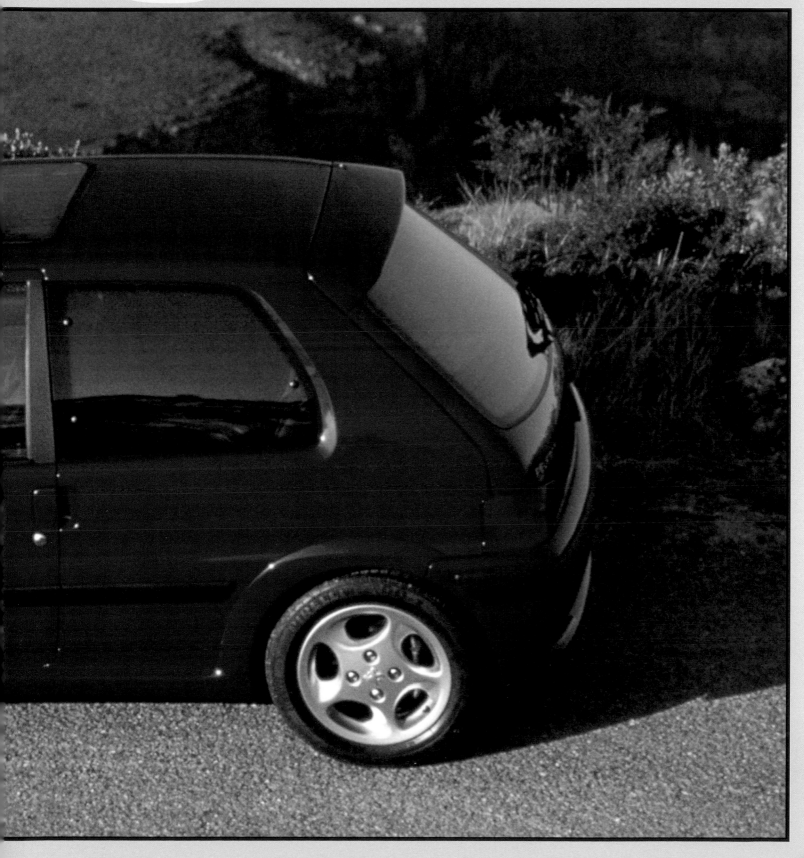

Plymouth Hemi 'Cuda

In the early 1970s custom car magazines were all the rage. It wasn't just the semi-naked women lolling suggestively on the bonnets of exotic American cars which attracted the attention of Britain's pimply youths, however. Oh no, they were busy putting go-faster stripes on their Vauxhall Vivas in a pathetic attempt to make it look like a 1970 Plymouth Barracuda. The 'Cuda is probably the prettiest muscle car to come out of the US and the convertible was a stunner. It came in groovy colours like lime green and putrid purple, and would guarantee as many babes in your passenger seat as you imagined you could handle cruising on stupidly hot nights in middle America – or the Purley Way.

Bruce Springsteen made things worse for the nation's youth by singing about 'hemi powered drones' and 'girls checking their hair in rear-view mirrors' on Born To Run. This ensured that the 'Cuda with the 426 Hemi – Chrysler's biggest big block producing a massive 425 hp – was the hippest, coolest car on the planet for about three weeks during the summer of 1973. But perhaps best of all, the Hemi 'Cuda came with a 'shaker hood' – an air scoop bolted directly to the carb and poking through the hood (bonnet to us), so that when the monstrous Hemi rumbled, the scoop could be seen shaking, ready to suck in any small children that strayed too close.

And the odd thing is, with that shaker hood, scantily-clad women couldn't lounge on the bonnet. **jc**

> *When the monstrous Hemi rumbled, the scoop could be seen shaking, ready to suck in any small children that strayed too close*

This is the car that Warren Oates drives across America in the film Two Lane Blacktop, playing tag with James Taylor and his hippy friends (one male, one female) in their '55 Chevvy. Oates, a distinctly unglamorous American character actor who looked like he knew how to drink and drive, spends a lot of time in the film raving about his car – a gleaming yellow hardtop with a black vinyl roof.

The cars in Two Lane Blacktop are undoubtedly the stars, and Oates' GTO is the principal lead. The GTO is clearly seen as representing the American automobile's future, racing against its glorious past, in the form of the not quite finished customised fifteen-year old Chevvy being driven by men with long hair who refuse to fight for their country, wash or get a job. Oates chews a lot of gum, snorts illegal substances and is convinced that the hippies are racing him. In fact, they just happen to be going in his direction. At one point in the film Oates persuades the hippy girl to get into his car, where he tries to seduce her with macho Army tales. So, Warren Oates drinks, smokes, takes drugs and drives too fast in a big, 366 hp powered coupe of which he's very proud and thinks that male hippies should be shot. The only minus point in his character is that he wears open-backed driving gloves.

Which is why the 1970 Pontiac GTO is in this book.**jc**

Porsche 944

No don't turn the page; it's not here. And nor would it be here even if this was a list of my top 500 cars. I'm afraid I don't like the 911. While I don't agree with PJ O'Rourke who described it as 'an ass-engined Nazi slot car', I find it an unpleasing shape and a triumph of engineering over design. However, I do respect those who like it, and after a run in an early 2.7-litre RS, I think I can see the appeal. But sorry, my favourite Porsche is the 944 Turbo. I fell in love with it in the days before they built the M40 and the quickest way between Britain's two biggest cities was up what amounted to a farm track. There'd be a lorry and then a tailback of ten cars, all trundling along at 35 mph, but realistically, there was no way past. And even if you did scrape by, you'd be in another tailback three minutes later. So it wasn't worth the bother or the risk. However, the 944 Turbo had such colossal mid-range grunt that it could flash past a hundred cars in one neck-snapping second. The figures aren't much to write home about but in real life, this was a stormer. With its gearbox at the back, there was perfect weight distribution too. And that, in turn, meant it handled well. And inside, the dash was solid, and sensible unlike that kid's party game you get in a 911. And it was fronted by a near-vertical steering wheel which made the driving position spot-on.

You'd have to search with a very, very powerful microscope indeed to find anything wrong with what was one of Germany's finest automotive hours. Whereas you can spot the mistakes with the 911, with the naked eye, from 40 yards.**jc**

I'm afraid I don't like the 911

Porsche 928S

A word of warning: when the Porsche's on-board computer says you have enough fuel left to do 17 miles be very careful because you, in fact, have no fuel left at all. And then you'll come close to being late for your father's funeral. Yes, it did run out and yes, a passer-by did call me a w****r as I trudged off in the rain for some more.

But the 928 – any 928, but the later ones are best – is a great car nevertheless. I know the ride was too firm for some tastes and that to purists, a front-engined, water-cooled Porsche is as wrong as a neon bible. But I liked the relentless grunt from its V8 motor and I respected its ability to be a supercar and a hatchback all at the same time. Yes it did 180 mph but moments later, it could be in a car park somewhere, not over heating and being as good as gold. You can buy 928s these days for less than the price of a used Mondeo and I suggest you do. **jc**

neither is this

Engine	V8, 4474–5400cc
BHP	219–350
Brakes	discs all round
Top speed	170 mph
0–60mph	5.7 seconds
When?	1978–1995
Today's cost	£8–£40,000

You can buy 928s these days for less than the price of a used Mondeo and I suggest you do

Range Rover

When I first saw the new Range Rover, I had to be brought round with smelling salts. What the hell have they done, I wondered, as I crashed to the floor in a dead faint.

Months later, my initial reaction was reinforced on a gloomy winter's evening, when I tried to hail one on London's Regent Street. I suspected it to start with, and I knew it for sure right then:

paint this brand new 4.6-litre status symbol black and it would look exactly like a Metrocab.

And then reports started filtering back to the office. Not only did the Range Rover look all wrong but it was about as reliable as a South American dictatorship. And to make things worse, there was no common thread – gearboxes, leaks, faulty electrics, you

name it, and the Range Rover had it. It was a walking automotive medical dictionary. I borrowed one last time I was in Dubai and in the space of four days, the CD jammed and the sunroof refused to close. Not much you might argue but this was a £50,000 car. I didn't want not much. I wanted nothing at all.

I certainly didn't want what happened on the fifth day. It caught fire and while

Range Rover captured in shock off-road moment. Must be lost.

Range Rover

parked on the hard shoulder, was rammed by a lorry.

I came home and washed the Range Rover from my mind. As far as I was concerned, it was just a joke and the important thing was to find a new car that could pick up its baton and become the best 4x4 in the world. After much careful deliberation, I decided to go for Jeep's Grand Cherokee but now, I'm not so sure again.

You see, I've recently taken up shooting and in order to get out there, people use 4X4s, and I mean, really use them. And there is no doubt that the Range Rover really will go where the Grand Cherokee, and the Explorer and the Shogun will not. I know owners hate the air suspension system but I discovered that the ability to change the ride height was an enormous help on deeply rutted tracks. I liked the wheel articulation too so that no matter how elaborate the terrain, the RR will keep all four tyres in contact with terra firma. And a tyre that's on the ground will have grip whereas one that's thrashing around in the sky, will not.

Then there's the engine. Forget the hopeless diesel but do not ignore the 4.6-litre V8 which has so much poke. Long after the feeble sixes found in the rivals have given up, the Range Rover is still out there, ripping up trees and putting out fires. And then, when it crashes from the wilderness, and back onto the road, it does a very passable job of becoming a car. I still wouldn't want to make a wild evasive manoeuvre on the motorway but I wouldn't want to do that in any high riding 4x4 ... in the same way that I wouldn't want to take a Porsche across a ploughed field.

No, accepting the obvious limitations, the Range Rover rides well, is quiet and has reasonably direct steering. And it is beautifully appointed as well. Wood and leather is absurd in a great many cars but not this one.

There are toys too like cruise control, and air conditioning, and a rear view mirror which dips automatically when the berk behind forgets to switch his full beam headlights off.

But best of all, there's that high driving position. You get this, obviously, when you're in any real off-roader, and it does let you sneer at other road users. But in a Range Rover, the sneer is somehow a little more genuine. Your car is not only taller than their's but it was almost certainly more expensive too.

So go right ahead, give it the full Elvis lip curling treatment, and be prepared to turn into a field if things get nasty. The Range Rover may not be built to high enough standards but there's no doubt that the design is right. It really can reach the parts that ordinary cars cannot. It is the best 4x4 by far.

But hold on a minute, you're thinking. It still looks like a Metrocab, and that is inexcusable. Surely, the old Range Rover which had all the same attributes is the better bet?

Well here's a funny thing. I no longer think the old car does look better and call me fickle, but I no longer think the new one looks like a taxi. The dark green one looks utterly wonderful.

So get one, get out there, and kill some wildlife.**jc**

Reliant Scimitar SE5

Yes, I know that Princess Anne drove one, and that Reliant are famous for their three-wheeled milkfloat the Robin, but this is different. Believe me.

Having been developed from the ugly but promising Reliant Coupe of the early 1960s, the Scimitar was the first fibreglass-bodied car which managed to combine practicality – it had a hatchback and revolutionary, individually folding rear seats – and performance. By cramming a 150hp Ford 3.0-litre, straight-six engine under the plastic bonnet, tacking on a fifth gear (quaintly dubbed an overdrive and worked via a long, plastic stalk which stuck out from the dashboard dangerously close to the driver's left hand) the Scimitar could get to 60 in under 9 seconds. And easily

manage a top speed of 120mph.

However, being very light and a rear-wheel drive, the Scimitar would simply not go around the longest corner in the wet without the back end swinging as wide as a barn door. Which is undoubtedly why Princess Anne and her horsey lot used to put bales of hay in the boot.

Other drawbacks of the daftly named by-Royal-appointment "estate" car, was the fact that Ogle's shapely designed front end meant that a small radiator had to be used, and whenever the sun shone for more than fifteen minutes a day, the car would overheat and the fake-leather interior would get really sticky. And you couldn't wear shorts to drive the thing because of the heat generated by the

engine which rested on top of your knees. The body wouldn't ever rot, of course, but it also wouldn't withstand an impact greater than that of the smallest pebble. But if it was good enough for Princess Anne, who are we to argue? Volvo certainly liked the design enough to base their estate version of the Saint's P1800 on it. **jc**

---

the princess

Engine..........	Ford V6, 2994cc
BHP	150
Brakes..........	discs all round
Top speed	125 mph
0–60mph	8.5 seconds
Big fan...........	Princess Anne
When?	1968–1975
Today's cost.................	£3,500

Renault A110

Technically this should be the Alpine-Renault, since that was how it always used to be written when, during the late 1950s and early 1960s these small, fibreglass bodied, rear-engined cars used to win every single mountain rally they entered. Or nearly.

Despite using some humble Renault running gear, the A110 was fast and extremely agile. In later years, after Renault bought the company from founder Jean Redele, they traded on Alpine's racing history to sell a V6 turbo-charged variation to the Americans. This is now known as the Renault GTA, and still bears some physical resemblance to the original Alpines, but is nowhere near as pretty. And it will go into a spin if you turn your head too quickly.

Even if you've never seen one on the road, you can tell the A110 was a great driver's car from all the pictures of them taking corners, either with wheels off the road, or pointing in the opposite direction to the car itself.

That the model was built for fifteen years with increasingly larger engines suggests that it wasn't as fragile as many critics suggested. Current prices suggest that the first, 1962-65 1000cc version was good, but the 1969-73, 1565cc version is the most sought-after, and will set you back more than £30,000. But it's probably worth it. **jc**

Renault Clio Williams

For year after relentless year, Renault invested millions in its Formula One programme, endlessly developing that incredible V10 engine so that the combined might of Mercedes and Porsche and Ferrari were left exhausted in its wake. But in the real world, they were selling cars like the Laguna which has about as much sporting appeal as afternoon tea. Renault never capitalised on its success in F1, marketing its cars to us on the strength of price and that infernal Nicole woman. But then came the Clio Williams, a mildly tweaked hot hatch which was most noticeable for its gold coloured wheels.

It was a limited edition special and the promise of good residual value coupled with the strength of the Williams name, made it an instant hit. So much of a hit, in fact, that Renault decided it wouldn't be a limited edition special after all and that anyone could have one. The people who'd bought on the first batch were thrilled. Not. But they should have been because whether there's one or 20,000 examples on the road, it was still a damn good car. Tight. Fast. Neat. Precise. And most of all, big, big fun.**jc**

Rover 3.5 litre coupe

This was the ultimate bank manager's car, as much a part of solid middle England as Captain Mainwaring. Mrs Thatcher used one on her first day as Prime Minister. I, on the other hand, used one as a banger racer in a late night stubble field race where anyone suspected of being sober was taken into the pits and breathalysed. If you were under the limit, you were forced to drink two pints before being allowed back.

And guess who won? I did, for while the Rover wasn't the fastest car there, or the most nimble, it absorbed damage like a sponge.**jc**

As much a part of solid middle England as Captain Mainwaring

Rover SDi Vitesse V8

The Rover SDi didn't really work terribly well in the rain but this had nothing to do with a surfeit of power over grip. No, it didn't like the actual rain drops because they would either break bits off it or cause the body to suddenly oxidise.

A slight shower and you'd be left holding a steering wheel in a pile of reddy brown dust. A hail storm and the whole car would be flattened. This is a shame because the SDi was a damnably clever piece of design. The interior had a cosy, snuggy, study-type feel in which a log fire would have been more in keeping than a heater. The exterior was even better. The front indicators had been shamelessly copied from a Ferrari Daytona but who cares? They made this car look, quite simply, sensational – so good in fact that in 1977 my grandfather sold his Bentley and bought one ... And another car as well to use when it was raining, or a bit foggy or too sunny, as just about any climatic condition you care to mention was enough to stop the Rover dead in its tracks. There were puny engined versions which were all right in a Marks and Spencerish type way but the 3.5 V8 was a real laugh. Had communism not settled in and around the factories where this car was being made (or not being made depending on the weather and whether Red Robbo had had a long enough tea break) we should be in no doubt that Rover today would have been strong enough to have bought BMW.**jc**

The interior had a cosy, snuggy, study-type feel where a log fire would have been more in keeping than a heater

Saab 99

Saab and BMW can argue until the cows come home about which one was first to fit a turbocharger to a mainstream production car but the answer is neither of them. It was Chevrolet in the 1960s in case you're interested. But who cares about records. I've always had a soft spot for Saab who, stuck up there in winter wonderland, don't conform. They were doing two-stroke engines and sticking two three-cylinder engines in cars while we were discovering bananas. And then, in 1967, came the 99 which was a curiously handsome car, even though it looked like nothing on earth. The big bumpers hinted at safety while the name with all those jet fighter connotations suggested plenty of power. But the oomph didn't really come until 1977 with the turbo. Suddenly the odd car had 145 bhp with the possibility for those who wanted more, to go to 210 bhp. Madness. Marvellous.**jc**

Sunbeam Lotus

This would have been the world's first hot hatchback were it not for one little drawback. Back then, the hot hatchback hadn't been invented. It was hot though. Years before the Golf GTi came along and set everyone's trousers on fire, this Lotus-propelled buzz bomb could do 0 to 60 in just 6.6 seconds. It was rear-wheel drive too so it could handle the power and the road at the same time. But it is not here because of its dynamic abilities and it certainly isn't here because it set new standards in terms of reliability. No it's here,

because it was the first car I ever crashed into. I remember thinking as I crested the brow in my Scirocco that there was a Lotus parked on the other side. I remember seeing it too, and thinking I'd still be able to miss it. But this was a stubble field, at 3am, and I may have had a few drinks earlier in the night, so instead of missing it, I ploughed a furrow right down its rear wing. I learned two lessons that night. First, that the Lotus was a fine-handling, ferociously fast car. And second, that you should never drink and drive.**jc**

It's here, because it was the first car I ever crashed into

Sunbeam Tiger

Sunbeam very nicely stuck a Ford 289 V8 under the bonnet of what was basically the wife's soft-top runaround

WELSH COUNTIES CAR CLUB

oh, I say

Engine	Ford V8/4737cc
BHP....................................	200
Brakes..............	discs/drums
Top speed	122 mph
0–60mph	7.5 seconds
When?	1964–1967
How many?	7066
Today's cost	£13–£16,000

Back in the early 1960s, there was still an air of adventure and optimism among men who had fought the Hun and flown astoundingly fast aeroplanes astonishingly close to the ground. They wanted a car that could recapture the thrill of shaking loose their handlebar moustaches and letting their white scarves flutter in the wind. So Sunbeam very nicely stuck a Ford 289 V8 under the bonnet of what was basically the wife's soft-top runaround, and didn't bother upgrading the brakes. Terry-Thomas, Leslie Phillips and Ian Carmichael loved it.

So did the Americans, where most of the cars were sold and corners don't exist. Which was odd, because it used the floorplan of the little, old, mild-mannered Hillman Husky and looked very pretty.

Of course, being made in Britain, the few that remained in the UK either fell apart with rust, or ended up wrapped around a lamp post which had been mistaken for a landing strip light. If you can find one though, drive it once with its original brake system, and then fit discs all round. **jc**

Toyota 2000GT

risen sun

Engine..	dohc, six cyl, 1980cc
BHP	150
Brakes	discs all round
Top speed	130 mph
0–60mph	9 seconds
When?	1965–1970
How many?	337

For 30 years Toyota had been busying itself making nasty white goods with wheels which is why the 2000GT was such a shock. Here was a car which had been styled by a German count, which had the same type of chassis as a Lotus Elan and which had a 2.0-litre twin cam engine. 0 to 60 was dealt with in 8 seconds and it could hit 137 mph but it was the handling that made all the waves. A Toyota that was fun to drive? Even today, that's unusual. However, it was not cheap. Indeed, it cost more than a Porsche 911 and a lot more than an E-Type Jaguar so sales were sluggish. Even when they made a special convertible version for Bond in You Only Live Twice, the world remained uninterested. Me though, I've always been very interested indeed ... in how they got the 2000 so right and pretty well everything since so spectacularly wrong.**jc**

A Toyota that was fun to drive? Even today, that's unusual

Triumph TR5

When some bearded bloke in a chunky jumper with Triumph badges starts banging on about how the TR6 was the ultimate model, tell him to get a shave and shut up. Yes the TR5 was basically the TR4 with a six-cylinder, fuel-injected engine that offers much the same performance as the TR6, but there's one very important factor Mr Beardy bloke is forgetting. The TR6 looked as if a German with little time to spare simply chopped off the interesting curves in a half-hearted attempt to make the car look macho. Which is what actually happened. Karmann were given the job of tarting up the TR5 because Triumph's house designer was working on other things. But since most of what the company were to turn out had the elegance of a wet cardboard box, it wouldn't have made any difference if he had worked on the TR.

No, for my money (and that's around £10,000 these days), the TR5 PI (petrol injection) is the best-looking 150HP model of the lot. Beware, however, the TR250 of the same time, 1967-68. It might look like the TR5, but is powered by the 104HP four-cylinder engine and is nowhere near as much fun.

What I particularly like about the car is that it kept the carburettor clearance bonnet bulge of the TR4, even though the 5 had none. Which is probably just the sort of thing that annoys Mr Beardy. **jc**

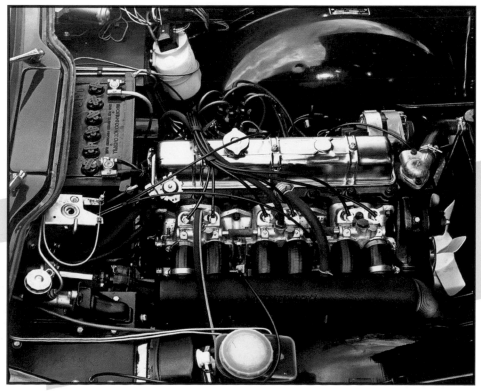

sod off beardy

Engine	2498cc
BHP	150
Brakes	front/rear drums
Top speed	120 mph
0–60mph	8.8 seconds
When?	1967–1968
How many?	2947
Today's cost	£13,000

Triumph Dolomite Sprint

The most important thing to know about the Triumph Dolomite is that it was designed so that a man could safely sit in the driving seat wearing his Trilby. By the time they came to put a twin-cam racing engine and sports wheels on the thing, that hat room was just as important. Only it wasn't for a Trilby, rather it was for the voluminous locks of the bare-chested lads who screamed around town without a seatbelt on, looking for likely lasses to accompany them on their nocturnal jaunts cross-country with fog lamps blaring. The Dolly Sprint was invariably yellow with a black vinyl roof, had a wooden dashboard and snazzy rubberised steering wheel. In the mid 1970s, the Dolly Sprint was the Number One fluff magnet. It was expensive you see, and offered more creature comforts than a Capri and a much more exhilarating drive than an Escort. It said 'this boy's dad is probably fairly well-off, they probably have a drive.' It also said that you could get your hair done on a Saturday afternoon and not have to worry about cramming it into a tiny coupe that night. Very important, that. **jc**

Triumph TR7 V8

When a famous Italian designer first saw the Triumph TR7 at the Geneva motor show, he stood in silence for quite some time, absorbing the profile of this new sports car. British Leyland waited with bated breath for his pronouncement but after a full five minutes, he had said nothing. Then, he walked round the back of the car and emerged on the other side where he stopped dead in his tracks. The crowd fell into silence waiting for the great man's thoughts. They came quickly. 'Oh my God,' he said. 'It's the same on this side as well.' In a couple of seconds, he had said it all.

After a long line of glorious Triumph roadsters, hopes were high for the TR7 but BL had dashed our aspirations on the rocks. First, it wasn't a roadster because at the time, it seemed likely that open cars would be banned in the USA. And second it was hideous, like a 12th century gargoyle, only worse. And third, it was built with the precision and care that normally goes into the public lavatories at a point-to-point. If ever I get round to doing a book on the worst cars ever made, the TR7 will be right up there in the top ten. But the roadster which emerged when American busy bodies decided not to outlaw convertibles, was actually quite handsome. And some of these were sold with the 3.5-litre V8 and that, really was what the TR7 should have been in the first place. We have swooned over V8 TVRs for years now but the TR7 was actually the first sports car to benefit from this wonderful brute of a motor. It was so good it made us forget about the woeful build and suspect handling. In fact, drivers would still have a smile on their face as the car slewed into a telegraph pole because the chassis had rusted away and the hopeless rear suspension had caused the tyres to lose grip. In a TR7 V8 – the TR8 tag was a nick name – you died happy.**jc**

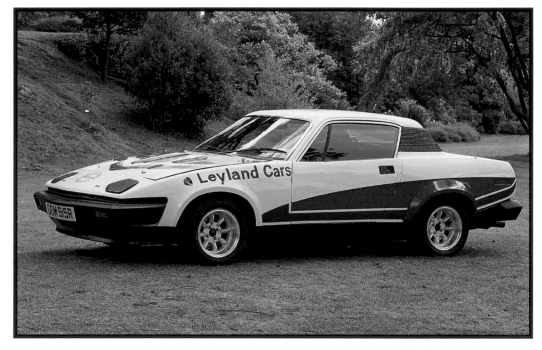

Purdy good

Engine .... Rover V8, 3528cc
BHP 133 (carb)/137 (injected)
Brakes .... discs/drums
Top speed .... 135 mph
0–60mph .... 8.5 seconds
When? 1980–1981
How many? 2497 (2308 in US)
Today's cost £7,500

TVR Cerbera

This is the only car I've ever driven where I almost never used full throttle

TVR was named after its founder TreVoR Wilkinson and, in its early days, made plastic kit cars. These were not good credentials. When I first became aware of these strange-looking creations, they had Ford engines and odd bodies which had about as much appeal as a stubbed toe. And then things got worse. In the mid-1980s, the Tasmin wandered onto the scene looking like it been designed on an Etch-a-Sketch. It was all straight lines and points, and frankly I didn't care that it was available with a 3.5-litre V8. It looked a little better, I'll admit, when the convertible was launched but it was still, basically, a kit car. And that meant heaps of trouble.

Now I know some people do enjoy cars that don't work properly. It means they have an excuse to spend a weekend on their backs, covered in grease with their fingers in holes that are much too small. I also like being on my back, covered in grease and playing with holes, but I prefer it if cars arc not involved.

However, in 1981 a chap called Peter Wheeler invested his considerable life savings in TVR and bought the company. Suddenly, the kit car started to growl and everyone started to take note. He went back to basics, making curvy two-seater convertibles which, on their own, would have been good enough to turn a few heads; but it wasn't the looks that were wooing the crowds. It was the noise. In a Porsche Boxster, there's momentary rasp from the exhaust at exactly 5,000 rpm which, I'll admit, sounds good. But they had to engineer that noise in. It didn't occur naturally.

Well no matter what TVR you happen to be talking about, no-one will hear you because they make noises that, really, belong in the jungle. It isn't engineered sound. It's just there: a rumble at low revs, a bellow in the mid ranges and a howl at the top. There's nothing refined about it either. There's none of the

majesty you get from an Alfa V6, say, but who gives a damn about majesty? You may be the most elegant dancer at the nightclub but if the bloke at the bar is hung like a baboon, he's going to get the girl. And believe me, if the TVR had a packet, it would be 19 feet long. Some cars are girly but not this one: it's a man, and not a washboard-stomached male model either. This is a big, hulking bouncer with a neck like a birthday cake and a fondness for biting the heads off dogs. Now until this point, I've been talking about TVR generically, but referring mainly to the Chimera and the Griffith which use versions of the Rover V8 and are traditional sports cars.

But when BMW bought Rover, Peter Wheeler said he wasn't having anything German in his cars and set about designing his own motor. Ordinarily, that would cost about a billion quid and take ten years but he had it up and running in a tenth of that time, and it cost about 30p. But it is one hell of an engine. As Wheeler says, Lexus won't be queuing up for its refinement but if it's blood and guts power you want, then it's fantastic.

With a flat plane crank, it shares much in common with today's Formula One racers, as does the car to which it is fitted –the Cerbera. I took this wheeled atom bomb to a runway and raced it against an Aston Martin Vantage, a Dodge Viper, a Caterham JPE, a Lotus Esprit V8 and a Porsche Turbo, to see which could get from rest to a point a mile away in the quickest time.

And it was a joke. The Porsche made a bit of a fight but the Cerbera just walked all over it, covering the distance, unbelievably, in under thirty seconds. Then the driver got out, got changed, had a bath, went to the pictures and was at home in bed before the rest bumbled across the line. This is not, however, an easy car to drive. Even getting it off the line quickly requires patience and

practice, and practice means you'd better have deep pockets to pay for all the tyres you'll wreck. Too many revs and the rear wheels just spin uselessly. Too few and you may as well be in a Metro. But get it right, and you'd better have strong neck muscles. The same things happens when you get to a corner. Too fast and you can forget all about electronic sensors or four-wheel drive to help you out. It's you versus the Cerbera and the prize is life.

There is so much power that you learn to have super control over your right foot. Think about it: in your car, if you press the throttle down by, say, an inch when you're already doing 60 mph, not much happens. Even if you have a BMW 328i. But if you press the accelerator down by an inch in a Cerbera that's dawdling along at 60, in the blink of a bat's eye, you're doing 100, and that's a worry if you happen to be cornering at the time. If you're a good driver, you learn to be careful in this car. And if you're not a good driver, being careful won't help. It'll rip your heart out.

I cheated by being careful to start with, resisting wherever possible using too much power. Seriously, this is the only car I've ever driven where I almost never used full throttle. Give me a Vantage and I'll go mad, give me a Cerbera and I'll be like a mouse creeping past the altar during evensong, while wearing carpet slippers. Or I'll be admiring the interior which, quite frankly, is in a class of its own. The boys at TVR took a look at the various switches and instruments needed and thought why should they go where everyone else puts them? So they didn't and that's why you will even find dials under the wheel. And why not? You'll note too there are no door handles. To get in, you press a button under the door mirror. It gives the car a clean, sleek look. A look you can enjoy right up to the moment when you hit a tree.**jc**

TVR Cerbera

woof woof

Engine.................... V8, 4185cc
BHP...................................... 350
Top speed .... 170 mph
0–60mph .... 4.6 seconds
When? 1994–present
Today's cost £40,000

VW Scirocco

Today, it's accepted that most affordable sports coupes have front-wheel drive. But when the Scirocco came along in the late 1970s, it was seen as a bit odd, a bit light on its feet, a bit AC/DC perhaps. Its chief rivals at the time were the Ford Escort RS2000, the Alfa GTV and the Capri. These were men's cars with prop shafts and basic, meat and two veg undersides. The Scirocco wasn't even handsome. It was pretty. It was a girl's car. But then came the GLi, and later on the GTi which had the same 1.6-litre, fuel-injected engine as the Golf GTi, and the same suspension and basic underpinnings. A powerful Scirocco had some appeal, which is why I bought one. I'd just left the north where a Scirocco ownership would have earned a fist in the face, and moved down south to experiment

with 'jus' and drinks' parties. And that car, let me tell you, was a serious bird puller. By today's standards, its brakes were hopeless and its steering was woefully heavy around town but this was of no moment. You only had to tell someone that you had one and you were in their bedroom. I guess that at the time, it was fun to drive too but quite frankly, I used it as a chat-up line rather than a car so I never really found out.

I did, however, once take it stubble field racing in the middle of the night but just before it, or I, had a chance to find out how good we were, I crashed. Or rather it did. I was too drunk to know what was going on. The best model was the Storm which came in green or brown and had leather seats. These were ideal, because you could wipe them down more easily. **jc**

VW Golf GTi

We all know the first man to break the sound barrier was Chuck Yeager, but hands up if you can name the second. We all know the first man to break the four-minute mile was Roger Bannister but does anyone out there have a clue who was next? I'm ashamed to say I can't even remember who followed Neil Armstrong down that ladder. There are many hot hatchbacks in this book, in the same way that there have been many great runners since Roger Bannister, but no-one will ever forget the name of the first – the Volkswagen Golf GTI. **jc**

VW Corrado VR6

After 54,000 miles in a Mark One Scirocco, I bought the Mark Two version without even bothering to take a test drive. This was a mistake. It was a mistake because Volkswagen had taken all the bad parts of the original car – cramped interior and heavy steering – and wrapped them up in a much uglier and heavier body. It didn't have enough power, even when the 1.8 came along, and it was about as much use for pulling women as having dog dirt on your shoe. I hated it. So, when it was replaced by the Corrado, I moved out of the VW camp. Which was another mistake because the Corrado was – is – a belter. The supercharged version was fun but the VR6 was about as close as you could get at the time to a £20,000 Ferrari. It sounded wonderful, looked gorgeous and went like stink. And VW had reminded themselves that quality was an issue too. Whereas my Mark Two Scirocco broke down all the time, the Corrado was bullet proof.

This means it'll last a while and that's good too because I have a sneaking feeling that it will be seen in years to come as a classic.**jc**

saintly

Engine	1986cc
BHP	120
Brakes	discs all round
Top speed	115 mph
0–60mph	9.7 seconds
When?	1971–1973
How many?	8078
Looks	Elegant Scimitar
Today's cost	£4,750

When Roger Moore was The Saint, he used to drive a Volvo coupe, the P1800. When Ian Ogilvy was The Saint, he used to drive a Jaguar XJS. This is clearly why The Saint was never the success that Bond was. The P1800 was relatively pretty when parked next to a Morris 1800, but unfortunately it wasn't much faster. Twenty-five years after Roger hung up his eyebrows as Simon Templar, a wet wimp named Michael in the US TV series Thirtysomething, drove a P1800 through various emotional crises, marital spats and New Man conflicts. He lived in a city which had lots of snow and rain, and the nights were long and dark. And that's exactly why this has made the list.

Here is a car that looks so good, women will get into it even with whining men who cry and eat quiche. Forget the fact that it can't go very fast, it looks as if it should. When you've either reached the age of worrying about your childrens' school fees but still want to pose in an unusual car, or want to let the nanny drive a safe (it is a proper Volvo) attractive car which can fit in the dog and shopping, try to find one of the Estate versions, the P1800ES.

Not for nothing is the P1800ES the most expensive classic Swedish car around. They never break down or rust, and people who own them love them so much they don't want to let them go.**jc**

Volvo 123GT

LPY158F

It's probably something to do with all the snow in Sweden. That and the long, dark nights. Low visibility and slippery roads dictated that Swedish saloon cars traditionally be safe, hulking, ugly lumps of rust-resistant metal which, try as they might, never manage to get the pulse racing. Even the Saab 94, with its twin-stroke engine, funny gearbox and remarkable rallying triumphs managed to look like the kind of beetle which was

going to survive any atomic explosion.

But then again, there's a lot to be said for automotive safety. Not that I know what that is, of course. What I do know, is that Volvos a) are never going to have boyracers attempting to shove them off B roads in Essex, and b) if by any strange chance a Volvo driver does ever lose control, the armour-plated body work will protect all occupants against an onslaught from the toughest obstacle in

its path. The Amazon 123GT makes it into this list because, despite weighing more than a Sherman tank, it does look better, and is just as safe. It also goes faster than the average Volvo of the time, and features a lot of chrome and extra foglamps.

Plus most importantly, 30 years after being made, there are still a lot of them around. How many Humber Sceptres do you see these days, eh?**jc**

ABC 258

Volvo was all about safety and now, thanks to the T5, it isn't

Volvo T5

To demonstrate that he was not cut out for action, Nick Cage speaking to Sean Connery in the film, The Rock, says 'I drive a Volvo'. And that was fine. We had the guy clocked. We all know what a modern Volvo driver looks like – tweedy with a hint of Hermes. We know what they drive like too. Badly, with a hint of disdain for anyone who dares to overtake their lumbering tank. But then came the T5 with its turbo motor and now, the whole damn world wants a slice of the action. One car completely changed the way we think. Volvo was all about safety and now, thanks to the T5, it isn't. That is the kind of trick which would leave Paul Daniels gasping. And that makes the T5 quite a car, especially if the said Mr Daniels never recovers.**jc**

motor movies

The Italian Job (1969)
Michael Caine trashes Aston Martins and E-Types, Mini Coopers outrun Police Lancias, Noel Coward gives the orders.

Dirty Mary, Crazy Larry (1974)
Peter Fonda and Susan George crash loads of American cars running away from the cops. Great stuff.

Two-Lane Blacktop (1971)
Warren Oates. What a driver, what a man. A brand new GTO, a fag, and the boot's full of Jack Daniels.

Duel (1971)
Dennis Weaver trying to get away from a bloody big truck whose driver he can't see, which is trying to crush him (obviously not a McCloud fan).